D1733780

Recycling

Recycling

Meeting the Challenge of the Trash Crisis

Alvin, Virginia and Robert Silverstein

G . P . PUTNAM'S SONS
New York

This book is printed on recycled paper.
Our thanks to Maarten de Kadt, Ph.D., formerly Research Associate, Solid Waste Research, INFORM, for reading the manuscript.

 G. P. Putnam's Sons, a division of The Putnam & Grosset Group, 200 Madison Avenue, New York, NY 10016.
Published simultaneously in Canada.
Printed in the United States of America.
Diagrams by Colleen Flis
Book design by Joy Taylor

Library of Congress Cataloging-in-Publication Data
Silverstein, Alvin.
Recycling: meeting the challenge of the trash crisis / Alvin, Virginia, and Robert Silverstein. p. cm.
Includes bibliographical references.
Summary: Discusses different methods of recycling waste, associated advantages and problems, and the possible future.
1. Recycling (Waste, etc.)—Juvenile literature. [1. Recycling (Waste) 2. Refuse and refuse disposal.] I. Silverstein, Virginia B. II. Silverstein, Robert A. III. Title.
TD794.5.S555 1991 363.72'82—dc20 91-12 CIP AC
ISBN 0-399-22190-5

10 9 8 7 6 5 4 3 2 1

First Impression

Photograph Credits
Aluminum Company of America page 63.
Environmental Action Coalition pages 20, 24–25, 34, 39, 44, 49, 65.
Susan Fowler-Gallagher pages 28–29, 56–57, 78–79.
General Electric Research and Development Center page 81.
Diagrams on pages 9, 11, 12, 14 and 16 adapted from *The Biological Sciences* by Alvin Silverstein, Rinehart Press, San Francisco.

Contents

1 • Spaceship Earth

THE GIANT spaceship speeds onward in its long voyage between stars. Its passengers and crew know that they will not reach port in their lifetime—or even in the lifetime of their children. There will not be a chance to stop off and pick up supplies, or to pull in for emergency repairs if something breaks down. What they brought with them will have to be enough, if they are to survive the trip. Not only tools and equipment but even air and water and food are very limited.

No spaceship could have room enough to pack all the supplies the travelers will need for a trip that will last for generations. But their life-support systems are very efficient. Tanks of algae grow under bright lights, taking in carbon dioxide from the stale air and supplying life-giving oxygen in exchange. Fast-growing fish graze on the algae and supply food for the voyagers; the algae, too, are harvested for their store of protein and other food materials.

The astronauts' body wastes are not thrown away. Instead, they are carefully saved and separated into things that can be reused—pure water for drinking and washing; fertilizer for the algae and the trays of growing plants. Worn clothing, broken tools, and other "useless" things are made useful again by mending or recovering and recycling their raw materials into new things. Nothing on the spaceship is wasted—the astronauts can't afford to waste anything!

The spaceship traveling between the stars is a familiar theme in science fiction stories. One day we may build such ships—small, self-sufficient worlds that can supply all the needs of their human passengers. Meanwhile, we are already traveling on just such a spaceship—our planet Earth.

Our planetary "spaceship" does not have an outer casing of metal. It is wrapped in a layer of gases, the atmosphere—which includes the air we breathe. Spaceship Earth is provisioned with a wealth of supplies, from minerals and fuels to water and oxygen. Not only humans but uncountable billions of other living things share the planet's resources.

So far, we humans have tended to act as though the earth's resources were limitless. We took what we wanted, used it, and if any unwanted bits were left over, we dumped them into any convenient place.

The ocean is vast, we thought. If we dump some loads of trash into it, they'll just disappear. But then we started to find the bodies of water birds, drowned because their feathers were coated with oil spilled from tankers, or entangled in the plastic rims from soft drink six-packs. Bathers found their beaches littered with syringes and other medical wastes.

Just pile the trash up in a hole in the ground, we said—somewhere out of sight. But the landfills spread, and then they were no longer out of sight (or smell), and they were leaking poisons into our water supply.

If a crew of astronauts took care of their spaceship that irresponsibly, they would never survive long enough to reach their destination. Can *we* survive if we continue to abuse our Spaceship Earth?

That was a question few people even thought of asking a generation ago. But now millions are becoming aware that we—and our planet—are in trouble. The results of our carelessness and that of past generations are starting to show up. It is getting harder to find new deposits of oil and minerals. Our waste products are piling up into mountains, and we are running out of places to put them. Some of our waste products are poisoning our water and air. These poisons can make people and animals sick, and they are killing trees and other plant life.

Chemical wastes in the air may even be changing Earth's atmosphere. More people are getting skin cancer these days, and medical experts believe our heedless ways are to blame: chemicals used in air conditioners, refrigerators, and some spray cans have reacted with a gas called "ozone" high up in the atmosphere and removed part of this protective ozone layer that used to filter out the harmful ultraviolet rays in sunlight.

Many scientists now believe that carbon dioxide, a gas sent into the atmosphere by forest fires, automobile engines, factory smokestacks, and home heaters, is acting like a blanket, holding in more of the heat from sunlight. This "greenhouse effect," they say, may result in great changes in our planet's climate. At some time in the future—perhaps within the next century—the ice at the poles may melt, causing the oceans to spill over onto the lands and flood many of our largest cities. The rich farmlands of today's world may turn into hot deserts.

Is there anything we can do to get Spaceship Earth running smoothly again? Or is it too late?

Most scientists believe that there is still time to save our Earth and make it a more pleasant place to live—if we act now. We have to learn to use our planet's resources more thriftily. Some say we should cut back on our use of energy and materials and live more simply. But that alone would not be enough. We also need to stop using Earth's resources *wastefully*. We need to make more use of renewable resources—like paper and wood, which can be replaced by growing more trees. Metals and the petroleum used to make plastics are nonrenewable resources. Our supplies of them are limited, and once we use them up, there will be no more. Many of the things we throw away today could be reused. What we think of as "trash" could be a rich resource instead of a problem. In fact, reusing our wastes or recycling them into new usable forms could help to solve many of the problems our careless ways have created.

All over the world, people are beginning to work at solving the trash crisis. Communities are setting up programs to collect various kinds of reusable throwaways, and ordinary people are more willing to put the extra effort into sorting their trash to make it easier to recycle. We are starting to make some progress, but there is still a long way to go.

In this book we'll find out more about the trash mess we have created and the dangers it poses to the passengers on Spaceship Earth. Some serious trash problems are the result of industrial wastes—the refuse generated in mining and manufacturing, for example. But we will focus mainly on the part of the trash problem that affects people most directly—the thrown-out food matter, waste paper, and other refuse of homes, schools, and offices. We'll explore some of the ways recycling

Speaking of Trash . . .

SPECIALISTS on environmental problems use the term *solid waste* for the useless things that are discarded by households, offices, industry, and agriculture. *Municipal solid waste* refers to the unwanted materials that are collected by public or private haulers from homes, commercial establishments (such as offices and restaurants), industry, and government agencies. It does not include the wastes generated by industrial processes—the slags from ore smelting, for example. (In the United States, industrial wastes account for close to 99 percent of the more than 10 billion tons of solid wastes generated each year.)

In everyday life, people use more familiar terms for society's throwaways: *garbage, refuse, rubbish,* or *trash.* Garbage sometimes refers only to food waste (orange peels, chop bones, and spoiled leftovers, for example); but often it is used simply as a synonym for trash, to refer to any unwanted or useless material.

Whether you call it solid waste, garbage, or trash, this category does not include another common type of waste material: *sewage*—the unwanted material that is flushed down the drains of homes and industry and thus can be regarded as a form of liquid waste (even if it started out as a solid or has its liquids removed during processing).

can help to solve the trash problem, some of the programs that have already been set up, some things that individuals can do to help, and some promising ideas for the future. But first let's take a look at the lessons our own planet has for us. Recycling may be a hot new trend, but it is also an old process that has been going on for millions of years.

2 • How Nature Recycles

Sunlight beams down on the meadow. The rays of sunlight have traveled millions of miles to reach the earth, but they still carry plenty of energy. Some of it warms the grasses and the soil and the water of the little stream flowing along the meadow's edge. Some of the sunlight energy is captured in the blades of grass and the leaves of the trees in the nearby forest. Like tiny factories, the leaves use the captured energy to power chemical reactions. Carbon dioxide, a gas from the air, and water drawn up from the soil through the plants' roots are combined to form sugars and other complicated compounds. These products will provide energy and building materials as the plants grow.

The meadow looks quiet and peaceful, but that is only on the surface. Look closer and you'll spot the grasses moving. A small brown mouse scampers up a tall grass stalk and reaches out to pull down its nodding head. Daintily the mouse plucks out the grass seeds and eats them, one by one. Now and then, with a little twitch of its tail, it deposits a small, dark-colored pellet on the ground—the remains of past meals, mixed with the mouse's body wastes.

A shadow passes. A hawk is circling overhead, lazily coasting on the air currents while its keen eyes scan the meadow below. It spots the movement in the swaying grasses and swoops down suddenly, with a thrust of its powerful wings and a harsh cry. When the hawk flaps up into the air again, its claws are gripping the mouse's body. Flying over to a tree at the edge of the forest, the hawk perches on a branch to take its meal. Bits of fur and other scraps drop down to the ground below.

A fly buzzes by. It smells the scent of the bloody scraps and lands on one of them to feed. A beetle scurrying along the ground finds another and bites off bits of it with its powerful jaws. Meanwhile, among the meadow grasses, beetles and millipedes are feeding on the mouse's droppings, finding food for themselves among the matter that the mouse's body could not use.

The meadow grows dark suddenly as a cloud moves overhead, blocking out the rays of sunlight. Thunder rumbles, and lightning crackles downward, striking the tree where the hawk had perched. The lightning bolt splits the tree trunk, and a large branch breaks off and crashes down to the ground. Rain pelts down on the forest and the meadow. Rainwater collects in puddles and drains down into the stream, carrying bits of soil, dead leaves, and other matter along with it. But soon the cloud passes, and the sunlit meadow seems peaceful again.

Scenes like this are repeated—with variations—every day, all over the world. On the African plains, lions stalk and kill antelopes. Buzzards, hyenas, and other animals feed on the lions' leavings, and beetles and other small creatures take their turn with the scraps. Microscopic bacteria and fungi grow on the bits of dead matter that remain, breaking it down further into simpler chemicals. Worms, too, take nourishment from the bits of rotting matter and help to distribute its chemicals through the soil.

In the ocean, microscopic plants in the surface waters take in sunlight energy and use it to grow. Shrimplike copepods and other tiny animals feed on the plants and are eaten, in turn, by fish and other sea animals. The animals' body wastes and bits of dead plant and animal matter drift down to the ocean bottom, where snails and other scavengers feed on them and bacteria help to break them down. Currents in the water bring the chemical products swirling up to the surface, where they help to nourish the water plants.

Food Chains and Webs

In the world of nature, plants—from tiny, microscopic algae to huge trees—are *producers,* using sunlight energy to power the chemical reactions that make food and materials that can be used to build and repair their bodies. The natural world has its *consumers,* too: animals that feed on the plants or on other animals. These animals and plants are linked together into *food chains.* A producer, such as grass, is eaten by a consumer—a rabbit, let us say. But the rabbit, in turn, may become food for a predator such as a fox or a hawk. Food chains can be very long: Tiny copepods, feeding on the microscopic algae in the ocean, may be eaten by small fish, which are eaten by larger fish still, and so on up the line until the largest fish in the chain is snapped up by a hungry whale. The chains can also be rather short: Some whales filter seawater through sievelike plates that catch and hold small sea animals—these huge whales can survive on a diet of nothing but copepods! With each step up a food chain, certain chemicals in the food tend to become more concentrated. So the effects of toxic pollutants on animals high up in the food chains—like dolphins and humans—may be very severe.

The food chains are not the whole story. They are linked together into complicated, interlocking *food webs.* The leaves of grass, for example, may provide food not only for rabbits but also for deer, grasshoppers, and many other animals, while mice nibble on the grass seeds and soil insects browse on the roots. Field mice can feed on seeds, nuts, tender leaves, fruits, and also some insects, spiders, and centipedes. The mice may be eaten by snakes, foxes, skunks, raccoons, bobcats, shrews, hawks, and owls, and each of these predators can also feed on other things if there are not enough mice around.

Both plants and animals produce waste products while they are alive, and when they die their bodies are another kind of waste. Imagine what a "trash problem" our planet would have if the bodies of all the ani-

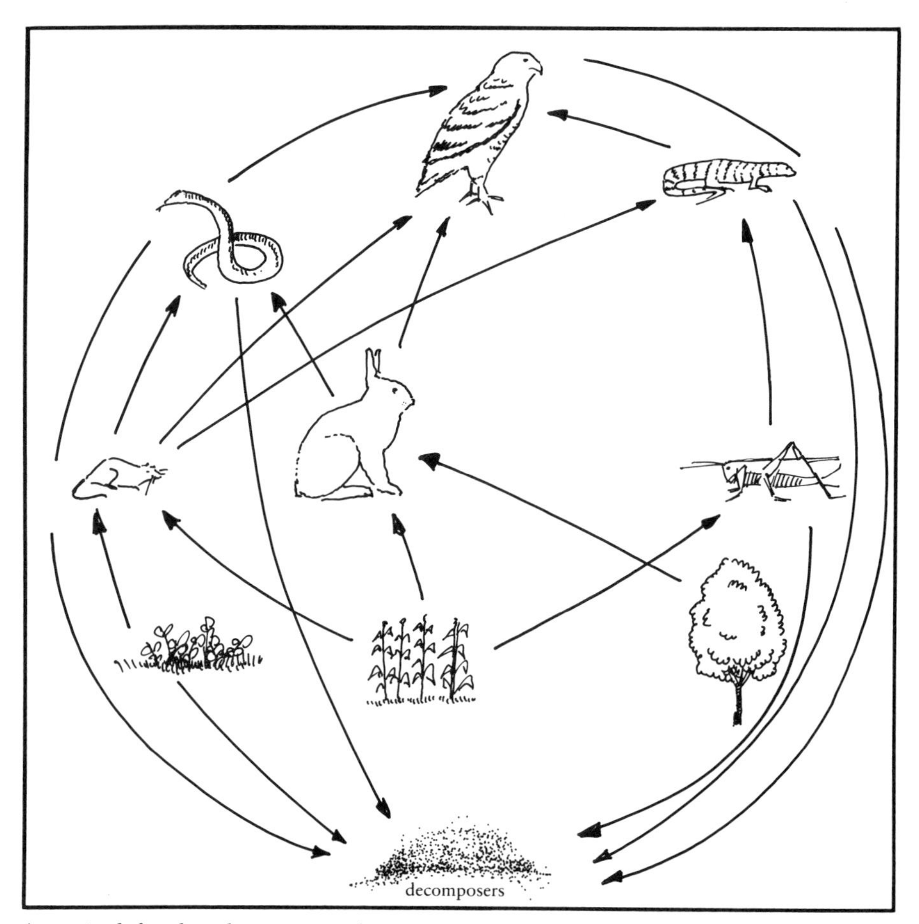

A typical food web in a meadow community.

mals and plants that died just lay where they fell! Dead grasses would choke out new plant growth, and the trunks of fallen trees would dam up streams and block the paths in the forests. Dead fish and other water creatures, piling up on the ocean bottom, would raise the water level and flood the coastal lands.

Of course, none of that could ever happen, because the world of nature includes not only plants that produce food, animals that feed on the plants, and animals that prey on other animals, but also a number of very efficient scavengers and decomposers. The scavengers range from buzzards and hyenas to beetles and millipedes. Together with bacteria, fungi, and other decomposers, they thriftily recycle the leftovers, converting them to forms that can be reused by the chains and webs of life.

Nature's Cycles

The resources of Spaceship Earth may be limited, but—so far, at least—the natural recycling processes have permitted one generation after another to reuse many of the same resources.

Take *water,* for example. Water evaporates from the surface of rivers and streams, lakes and oceans. The water vapor passes up into the atmosphere and forms clouds. When weather conditions are right, the water vapor in the clouds can condense into liquid water, which falls to the surface as rain. Some of the rainwater that falls on land areas drains off into streams, rivers, and eventually into the ocean.

Some of the rain is absorbed by the soil and drawn up through the roots of plants. Water pressure in the stems and leaves helps to keep plants sturdy—it is the only "skeleton" that some plants have. But plants continually lose water through tiny openings in their leaves, so they need a new supply. (If you forget to water a houseplant in a pot, it soon droops and wilts.)

Animals (including humans) are another important part of nature's water cycle. They drink water and also take in moisture in juicy fruits and leaves of plants. Your own body is really about two-thirds water. Animals lose some of their body water when they get rid of their body wastes, and there is water vapor in the breath they exhale. When an animal dies, the rest of its body water is returned to the cycle.

Nature also recycles important chemical elements. *Nitrogen,* for example, is a gas that makes up 78 percent of Earth's atmosphere. Most plants and animals cannot use pure nitrogen gas, yet this element is a part of many important body chemicals. A key link in the nitrogen cycle is played by specialized bacteria that live in the soil or in plant roots. These "nitrogen fixers" change nitrogen gas into chemical forms that plants can use. The plants further change these simple nitrogen compounds into proteins and other complex substances. Animals get

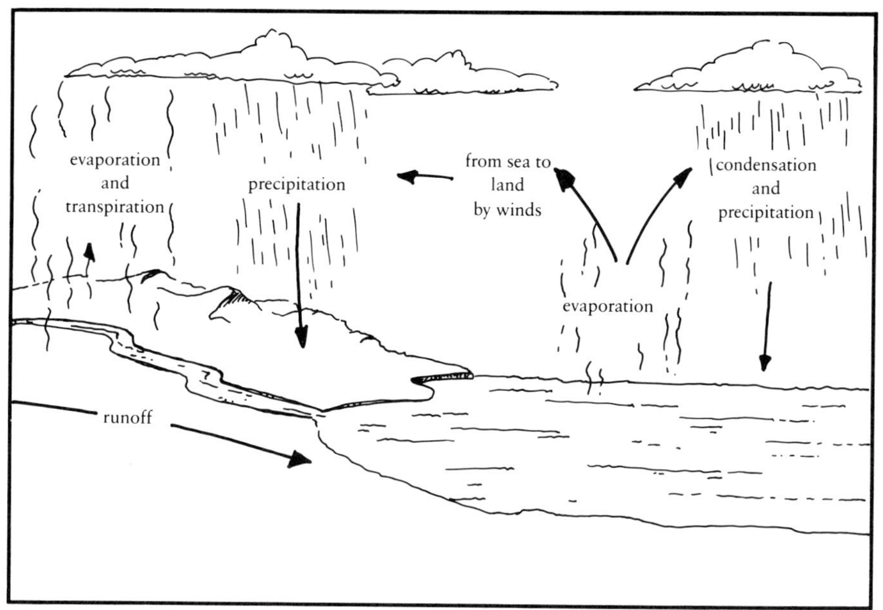

The water cycle.

the nitrogen compounds they need by eating plants or other animals.

When animals and plants die, decay bacteria break down the complex nitrogen compounds in their bodies into simpler substances, such as water and the strong-smelling gas ammonia. Other bacteria carry the process further to produce nitrogen gas. This is the reverse of the "nitrogen-fixing" chemical reactions, and it completes the cycle, returning nitrogen to the atmosphere.

The air over a single acre of land contains about 35,000 tons of nitrogen. The natural nitrogen cycle is perfectly balanced—the nitrogen removed from the soil by plants and animals is replaced by the nitrogen-fixing bacteria and the decay processes. But the crops grown by human farmers take more nitrogen out of the soil than the natural processes can put back. The nitrogen-rich body wastes of humans and farm animals usually do not go directly back into the soil. Instead, they are handled by waste disposal systems that may flush them into rivers and streams. There the extra nitrogen compounds can actually be deadly! They may produce sudden population explosions of algae that grow so thickly that they crowd the water surface. Then not enough oxygen

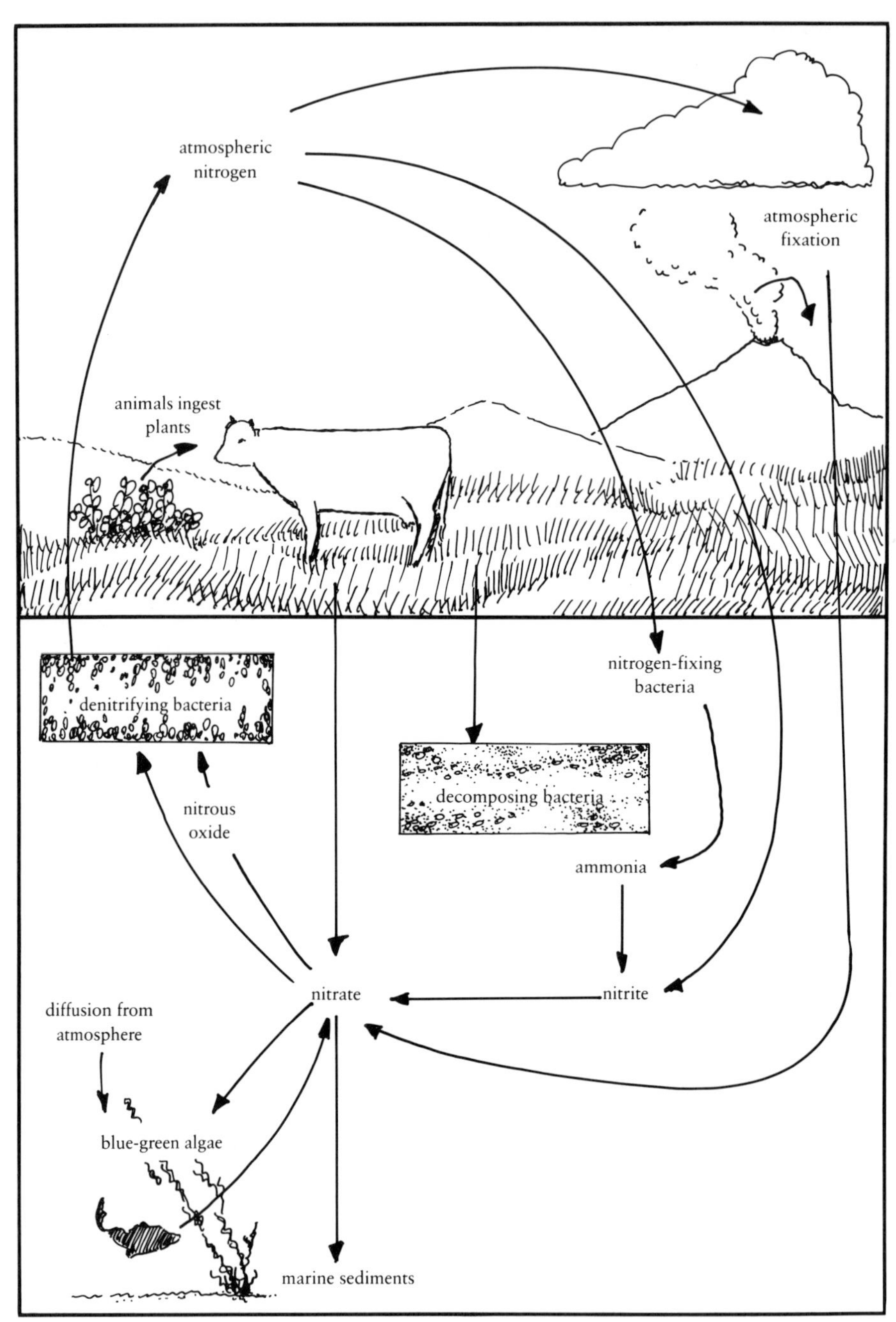

The nitrogen cycle.

from the air can enter the water, and fish and other water creatures choke and die. Meanwhile, the human farmers must add fertilizers to the soil, to replace the nitrogen the plants need for growth.

Human activities also put some nitrogen compounds into the atmosphere. When we burn fuels such as oil or coal, or even garbage, the nitrogen compounds they contain combine with oxygen from the air to form nitrogen oxides. These substances are irritating to breathe, and they can damage plants and even wear away the stone of buildings and statues. They are part of the troublesome "smog" that results from air pollution.

Living animals and plants play key roles in the recycling of another important element, *oxygen.* Although this gas makes up close to 21 percent of Earth's atmosphere, scientists believe that there was practically none of it in the air when our planet was first formed. At that time (billions of years ago), Earth's oxygen was mainly bound up in chemical compounds, in the water of the oceans and in rocks and minerals. For the first living things, oxygen was a deadly poison. But gradually creatures appeared that could make oxygen less poisonous or even use it to generate energy. Eventually there were plants that not only used oxygen but also produced it. In the chemical reactions called photosynthesis, they used sunlight energy to combine carbon dioxide and water into much more complex chemical compounds—food materials. During the reactions of photosynthesis, oxygen gas is released as a by-product. All of the oxygen in Earth's atmosphere was formed by plants. The microscopic algae that float on the ocean's surface in our present-day world, as well as trees, grasses, and other land plants, help to supply oxygen to the atmosphere.

Animals cannot produce oxygen, but they are heavy consumers of this important gas. They use it in respiration, a process somewhat like a burning fire, in which complex chemicals are combined with oxygen to release energy. Land animals obtain the oxygen they need by breathing air. Water dwellers use the oxygen dissolved in the water. Plants "breathe," too, but the amount of oxygen their respiration uses up is generally far smaller than the amount they produce in photosynthesis.

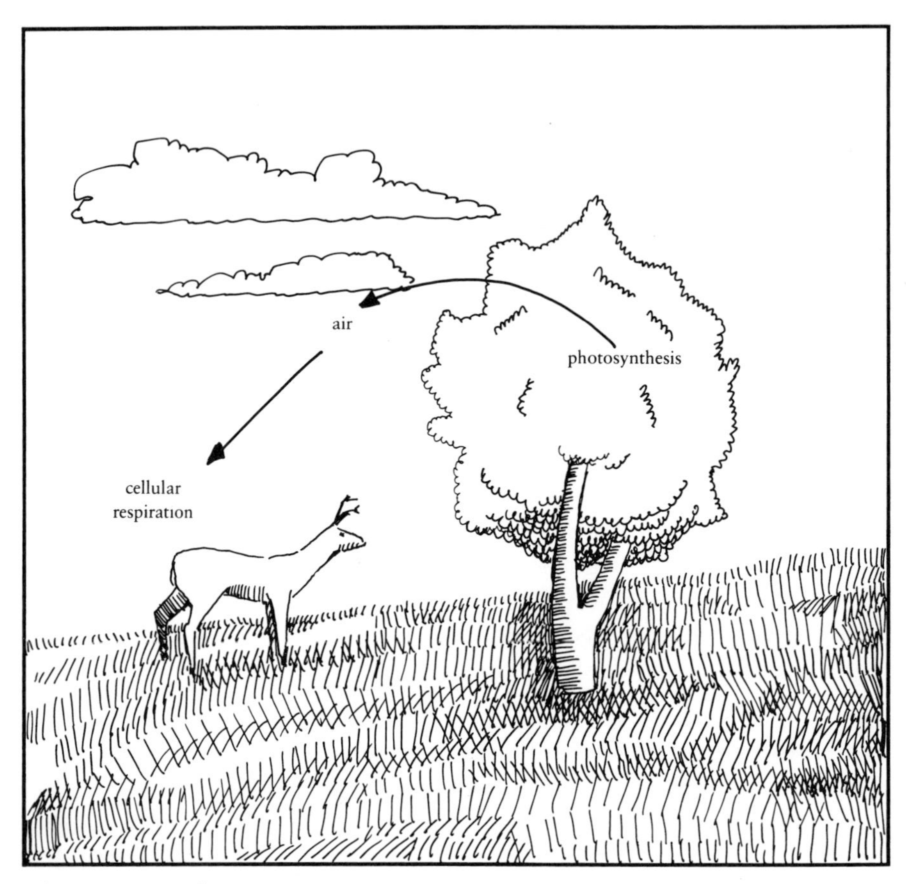

The oxygen cycle.

So in the natural oxygen cycle, plants add oxygen to the atmosphere, while animals use it up.

We humans use oxygen in another way, too, by burning fuels in the furnaces that heat our homes and power our factories, the stoves that cook our food, and the engines that run our cars. Fortunately, we haven't made a dent in Earth's oxygen supply yet—photosynthesizing plants release enough oxygen to make up for the amounts we use.

Nature's recycling may not be able to keep up with human meddling in another important chemical cycle, though. That is the *carbon* cycle, which is interlocked with the oxygen cycle. The connecting link between these two cycles is carbon dioxide, which is formed when the carbon compounds combine with oxygen—either in a burning fire or in respiration. Proteins, sugars, fats, and many other complex substances that are found in the bodies of plants and animals all contain carbon. This element is also a major part of the "fossil fuels": coal, oil, and gas. These are actually the remains of dead plants and animals, which were buried in the mud millions of years ago and "cooked" by the enormous temperatures and pressures deep below the surface.

There is not much carbon dioxide in Earth's atmosphere, compared to nitrogen and oxygen—only 0.035 percent. But this gas is very important for living things. It is the main raw material of photosynthesis—and thus the ultimate source of all food for our planet's creatures. The carbon dioxide in the atmosphere also helps to trap the energy from the sun, keeping our planet's temperature fairly constant, within a range that Earth's living things can tolerate. (That is the good side of the "greenhouse effect.")

Carbon dioxide from the atmosphere may be carried down into streams and rivers by the rain, or it may dissolve directly into the surface waters of lakes and oceans. (Earth's oceans contain sixty times as much carbon dioxide as there is in the atmosphere.) The tiny plants that float on the ocean surface in huge numbers use some of the dissolved carbon dioxide in photosynthesis and have a major share in producing oxygen gas. But some of these microscopic creatures use carbon dioxide to form hard protective shells around their delicate bodies. When they die, they sink to the bottom, taking some of the carbon out of circulation.

Animals and plants that build carbon into their body chemicals also take this element out of circulation—at least, temporarily. But when they die and their bodies are broken down, the carbon they contained is recycled. Carbon compounds in rocks may also be returned to circulation when they are dissolved out by rainwater or vaporized by erupting volcanos.

Many scientists are now worried that human activities are generating more carbon dioxide than the oceans and plants can soak up. They say

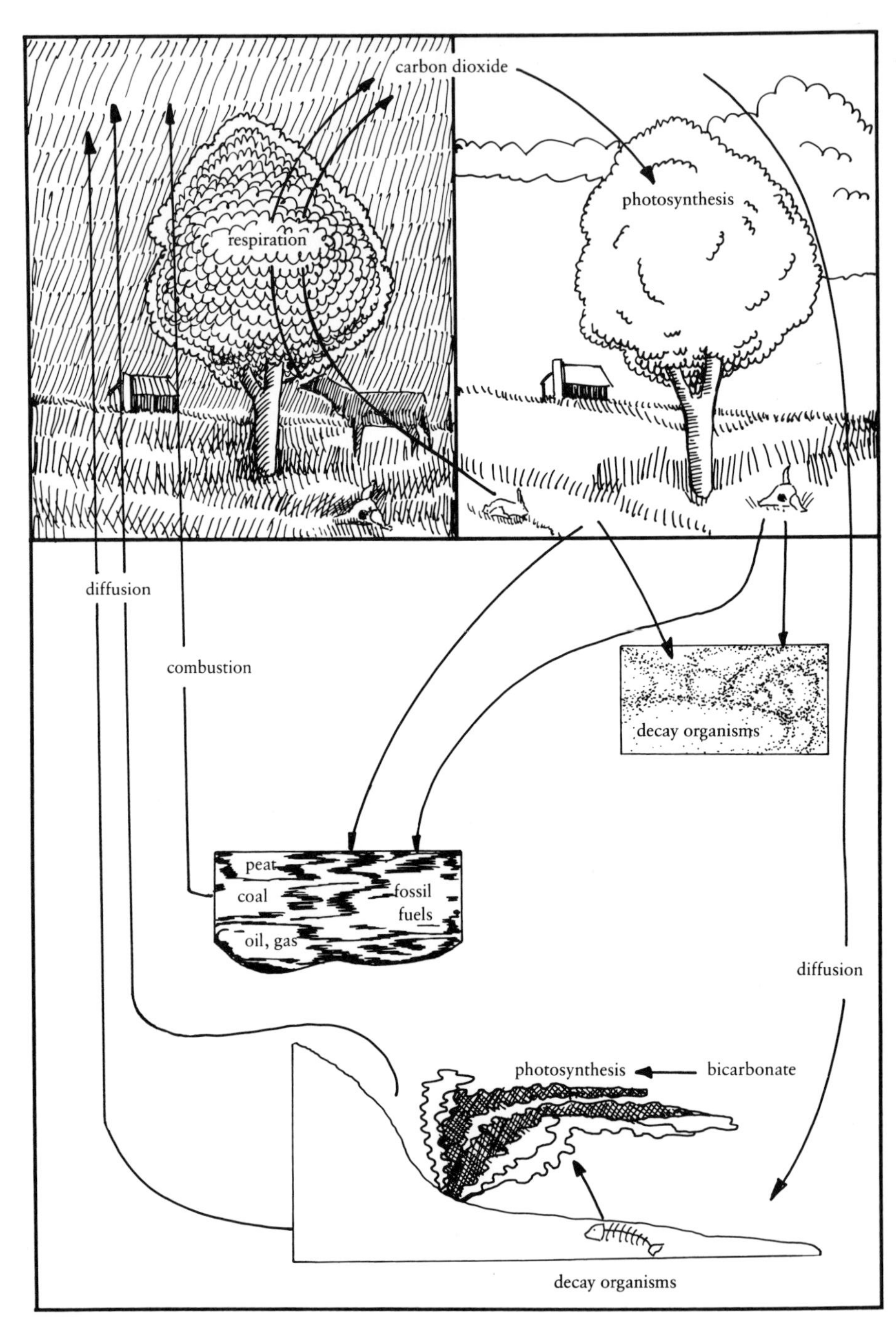

The carbon cycle.

the amount of carbon dioxide in the air is increasing, and that this rise is adding to the greenhouse effect, to produce a general warming of the whole world. No one knows yet whether the natural cycles will be able to keep things in balance.

New Problems

Human activities are causing other serious problems for our planet's environment, as well. The natural cycles are well equipped to handle the waste products of the normal activities of animals and plants. They could cope with human wastes, too, when there were fewer people around. But during the past couple of centuries, humans have been multiplying to enormous numbers, and the wastes that we generate are beginning to pile up in amounts that have no parallel in Earth's long history.

The United States alone, for example, generates more than 10 billion tons of solid waste each year. This amount includes about 180 million tons of municipal solid wastes (the "trash" from homes, offices, and small industries, which is collected and disposed of by their communities). Junked cars, appliances, and old building materials also contribute a substantial fraction. About 73 percent of these solid wastes are dumped in landfills, where each layer of trash is covered with soil and packed down with heavy bulldozers before a new layer is added. Landfills are usually started in a gully, swamp, or other natural depression, and are lined with layers of clay and synthetic materials to prevent chemicals from the trash from passing into nearby groundwater. They may be closed when the layers of trash and soil have built up to the level of the surrounding land. In some cases, though, the dumping continues as a literal mountain builds up. These mountains of trash are causing some special difficulties.

The natural decay processes, for example, usually involve the action of moisture, light, and bacteria. But moisture and light cannot penetrate into the depths of a landfill, and decay bacteria may not have enough air and other things they need to grow and multiply. A newspaper left outside in the yard will soon turn into a rain-soaked pulp, and it will not take long for its remains to decay. But when researchers dug into a landfill in Phoenix, Arizona, a few years ago, they found thirty-five-year-old newspapers in such good condition that their print could still be read!

Another problem with today's trash is that we have learned to make substances that never existed in nature. Working with the same chemical elements that make up complex natural compounds like proteins or starches, chemists have linked them together in new ways to form versatile new materials like polyethylene or polystyrene plastics. Some of these new materials are stronger, lighter, more durable, or cheaper than natural products. So the use of plastics is increasing. But when we throw these unnatural materials away, the natural recycling processes can't deal with them. Decay bacteria can't break their chemical bonds. A "Styrofoam" cup, buried in a landfill, could remain unchanged for hundreds of years.

Growing numbers of people are coming to the belief that we need to give nature a hand if we want to solve the trash mess. We need to develop more ways to reduce the wastes we generate and to recycle what is left; we need to apply them on a much wider scale. Recycling can accomplish two goals for the inhabitants of Spaceship Earth, conserving our valuable and limited natural resources and making a better life for ourselves and the other living creatures of our planet.

3 • The Trash Mess

DURING THE spring of 1987, the *Mobro,* a barge piled high with garbage collected in Islip, Long Island, wandered up and down the Atlantic coast. Each night, TV newscasts featured pictures of "the garbage barge" and updates on its forlorn search for someplace to dump its smelly cargo. Eventually the *Mobro* was unloaded in Brooklyn. The garbage was burned, and the ash was shipped back to Islip to be buried in a landfill. Meanwhile, the freighter *Pelicano* was zigzagging back and forth over the world's oceans, looking for a place to unload 28 million pounds of ash from the city incinerator in Philadelphia. At least eleven countries on three continents turned the *Pelicano* away during its long voyage. In November 1988—after twenty-two months at sea—the cargo was finally unloaded, but the owners of the freighter won't say where.

Many people have the idea that garbage and trash—the waste products of daily living—are things you can just throw away and forget about. That might have been true long ago, when Spaceship Earth was carrying far fewer people and had a lot more empty space.

Early humans had a simple approach to the problem of getting rid of trash. The remains of meals, broken dishes and tools, and other re-

A "garbage barge," loaded with solid waste, on its way to a dump.

fuse were simply tossed out the doorway. Archaeologists have found out a great deal about how these early people lived by studying the contents of such refuse piles, called kitchen middens. But a refuse pile can get rather smelly after a while; even worse, it attracts rats and other vermin and provides a breeding ground for disease germs. Some primitive people solved that problem by just packing up and moving away when the garbage became too much of a nuisance. Others developed more sanitary habits, burning their refuse or burying it in the ground.

As the human communities grew larger, trash collection services were organized. A common way of disposing of the communal wastes was to cart them out to an open dump at the edge of town. But as cities—and their dumps—grew, laws were passed to outlaw open dumps and establish "sanitary" landfills. (The dumps were not only smelly breeding grounds for vermin and disease but also serious fire hazards because of the gases produced by bacterial decay.) For a long time we were able to cope with our trash by burying it in landfills or burning it in incinerators.

Not anymore, though. All of our solid wastes have to go somewhere; yet, while we have more people producing more trash, we are rapidly running out of places to dump it.

About 73 percent of our garbage and trash is trucked to garbage dumps or landfills. But each month, landfills are closing—either because they are too full to take any more or because people in the area think they are too unpleasant or dangerous. Dumping garbage in the ocean used to be another option, but now it is forbidden by strict federal laws.

The amounts of solid wastes we produce are mind-boggling. Each year Americans throw away enough garbage to form a ten-foot-high mound covering an area big enough to hold close to 19,000 football fields! In 1988, the United States produced nearly 180 million tons of refuse—about four pounds per person, each day!

How did we get into this mess?

Our Throwaway Society

The first European settlers who came to America found an incredibly rich land. Vast forests provided timber to build houses, animals to hunt for food, and—after some of the trees were cleared—fertile soil for growing crops. To the west was the "frontier," beyond which stretched a seemingly endless vista of more forests, plentiful game, valuable mineral deposits, and huge empty spaces. (Actually, of course, the new lands were not really empty; native Americans were there, living for the most part in harmony with the land, taking no more from it than could readily be replaced by natural growth.)

When the United States became a nation, the "frontier" theme was an important part of its citizens' philosophy. They could freely enjoy the country's natural riches because there were plenty more, just waiting to be discovered and developed, out in the frontier lands. And with all those vast spaces, there was always somewhere out beyond the boundaries that inconvenient wastes could be dumped.

To be sure, another, equally strong attitude helped shape our philosophy. The early settlers might have had a wealth of natural resources, but they were very short of manufactured goods. If they needed things like needles and thread, hammers and nails, the settlers had to either make them themselves or import them—with long delays and at great expense—from across the ocean. So the traditions of hard work and thrift became American ideals. "Waste not, want not" was the motto. Outgrown clothes were handed down from one child to the next, and furniture was passed from one generation to another. If you received a package, you opened it carefully, neatly folding the paper to reuse the next time you had a package to wrap and adding the string that was tied around it to the tightly wound ball of string the family was saving. The waves of immigrants who came to the United States over the years had a similar philosophy. They had to work hard to establish themselves, and the small necessary things of life were very precious—to be used and reused carefully.

The development of industry brought many changes to our way of life. Manufactured goods became plentiful and cheap. Electronics brought an enormous variety of new products, and technological advances made them cheap enough for the average person to afford. As more people went to work in factories, there were fewer people available for services, and many of the old skills were lost. Now it is often hard to find someone to repair something that is broken, and it costs more to have a wristwatch or radio or toaster repaired than simply to throw it away and buy a new one! That seems crazy, but it makes sense economically: Most goods today are mass-produced. The materials and parts are bought in large numbers, which is cheaper per unit than buying single items, and they are put together by workers with limited skills, whose job may consist of nothing more than tightening a few bolts or running a machine that puts parts together. Repairing that same watch or radio or toaster takes someone with enough skill to figure out what is wrong, select the right replacement parts, and do all the operations of taking it apart and putting it back together—which may take as much time as it would take in the factory to make half a dozen or more new items.

Over the past century, the number of hours in the average person's work week has dropped, and there has been a boom in leisure activities; meanwhile, technology has produced television, video games, and other new things to use up our leisure time. The wealth of new goods to buy sparked an expansion of people's "wants," and soon one salary didn't seem to be enough for a family to get along on. So now, more than half of all married women work at outside jobs. Parents try to share the work of raising a family and doing the chores, while attempting to squeeze in some time for "fun," and they often feel pressured. "Convenience" products become important to them—sticking a ready-made dinner, complete with its own throwaway dish, into the microwave oven for a couple of minutes or taking the family out to a nearby "fast food" restaurant seems more sensible than spending hours shopping for fresh foods and cooking them from scratch.

Our economic system today emphasizes industry, and we measure our economic success in terms of "production" and "sales." There don't seem to be any columns for measuring "conserving" and "reusing." We spend billions of dollars on advertising to try to make people want and buy more goods. Much of this advertising comes in the form of "junk mail"—more than 12 *billion* mail-order catalogs are sent out each year! (Most of them promptly land in the wastebasket.) Few things are built to last—if something doesn't wear out, you won't buy a new one and the "production" and "sales" figures will suffer! Manufacturers of clothing, cars, and other goods that could last put an enormous amount of money and imagination into frequently changing

The landfill at Fresh Kills, Staten Island, New York.

695

styles. So someone with a perfectly good car or coat begins to feel uncomfortable with it because it is "out of style."

Packaging has come a long way from the simple paper and string that our great-grandparents used to save so carefully. Today's packages are designed for eye appeal, to make the goods look bigger and better, and to protect them during shipping and storage. Achieving those goals uses up huge amounts of paper and plastics—the package an item comes in often takes up much more space than the item itself. Most of these packages are not reusable. It would not make sense to save them, even if people had enough space to keep them. So we throw them away. The average American throws away about sixty pounds of plastic packaging each year! And that doesn't even count the paper bags and cardboard packaging.

All these developments have led to a huge amount of waste. We have lost the old philosophy of thriftiness. Most people today find it too difficult or inconvenient to save and reuse things. Unfortunately, many Americans have not lost the old "frontier" attitude, even though the lands that used to be frontiers are now crisscrossed by superhighways and dotted with housing developments and shopping malls. Until recently, few people have worried about where all our throwaways can go, and what effects they will produce on the environment. Now, though, the trash crisis is getting hard to ignore.

Mountains of Trash

The highest points in southern Florida today are the tops of landfills—literally, mountains of trash. The largest landfill in the world, located at Fresh Kills, Staten Island, greets visitors driving in from New Jersey to New York City with a powerful stench. Rising more than 150 feet into the air, with a volume estimated to pass that of the Great Pyramid of Egypt by the year 2000, the Fresh Kills landfill is continuously fed by hundreds of trucks and twenty-two barges each day, hauling 15,000 tons of fresh, smelly garbage. The flocks of seagulls that circle overhead, swooping down to pluck out juicy morsels, are content with the situation, but thousands of Staten Islanders have demanded that their borough secede from New York City, to escape being the dumping grounds for the other four boroughs' refuse. It is not only the smell that makes them angry; 2 million gallons of fluid gunk are leaking out of the landfill into the local water supplies and threatening the health of the local population.

What kind of garbage goes into the thousands of landfills in the United States? Researchers have actually sorted through the garbage collected in a variety of cities, counting, weighing, and measuring the different components. The largest part of the waste—about 40 percent by weight—is made up of paper products, with newspapers alone accounting for 12 percent and telephone books, junk mail, and paper packaging contributing substantial shares. (Americans get close to 2 million tons of junk mail each year—about 160 pounds for each man, woman, and child.) The next biggest source of trash (18 percent) is yard waste—raked-up leaves, tree branches, hedge clippings, and weeds. Metals account for about 9 percent; every three months, Americans throw out enough aluminum cans to rebuild the entire U.S. commercial air fleet! About 7 percent of our trash is made up of glass products, and another 7 percent is thrown-out food. People today are growing concerned about the use of throwaway "Styrofoam" containers and trays in fast food restaurants and cafeterias, but studies show that these objects account for only a quarter of a percent of the weight of all our trash, and plastics all together make up only 8 percent. (That statistic may be somewhat misleading, though. Polystyrene foam and other plastics are very light, and their proportion of the trash *volume* is much greater than their weight would suggest—20 percent of the total volume of solid wastes.) Disposable diapers make up a surprising 2 percent of the trash load. Other wastes, including wood, rubber, leather, and fabrics, account for the rest of the municipal solid wastes.

Toxic substances such as paint, used auto engine oil, and nail polish remover are not supposed to be thrown out with the regular trash. (They are supposed to be disposed of separately, in special leakproof facilities.) But most people don't bother to separate out an old paint

NOTICE

The incinerator in Duchess County, New York.
Is this a solution to the trash mess?

can, or a burned-out fluorescent bulb, or a dead battery, all of which contain poisonous chemicals that could leak out; and toxic materials make up about 1 percent of our trash load. With tons of toxic inks from newspapers and magazines, and decades of accumulated heavy metals from paints and dyes, the thousands of landfills in the United States are likely to be a source of dangerous substances for many decades to come.

MUNICIPAL SOLID WASTES IN THE UNITED STATES

	Generated weight	*Landfill volume*
Paper	40%	34%
Yard waste	18%	10%
Metals	9%	12%
Glass	7%	2%
Plastic	8%	20%
Food	7%	3%
Other	11%	19%

Note: Percentages indicated are for 1988, as cited in the 1990 EPA report, *Characterization of Municipal Solid Wastes in the United States.*

There are currently about six thousand landfills operating in the United States, but their number is shrinking each week. Some close because they are filled, others because they are found to be polluting local water supplies. As landfills close, communities are forced to look for other places to take their garbage—usually farther away. The trucking distances increase with each closed landfill, and the costs of garbage disposal go up. Proposals to open new landfill sites are typically greeted by local residents with a cry of "NIMBY"—"Not in my backyard!"

As landfills have closed, many communities have tried to solve the problem by building incinerators to burn their trash. But incinerators bring problems of their own. They are expensive to construct—which may mean higher taxes for the local community. They can be sources of dangerous pollution. For instance, the gases that incinerators send out into the air typically include various poisonous and potentially cancer-causing substances, such as dioxin. In addition, incinerators don't always burn the trash completely. Even in those that do, there is a residue of ash that contains heavy metals such as mercury and lead, as well as other toxic substances. So a community with an incinerator is still left with a trash problem—a very serious one, indeed. It is not very surprising that many communities are saying "NIMBY!" to proposals for new incinerators, too.

If landfills and incinerators just add to our problems, what can we do about the trash mess?

4 • Searching for Solutions

OUR SPACESHIP EARTH is a very complex world. Often, when we try to fix one problem, we introduce or worsen other problems.

For instance, since the energy crisis of the 1970s, the U.S. government has required that automobiles get more miles per gallon. The easiest way to do that is to make cars lighter in weight. (An engine doesn't have to work as hard to move the smaller weight along and thus burns less gasoline for each mile traveled.) Manufacturers decrease the weight of new cars by substituting lighter plastics for parts that used to be made of steel and other metals. Each year, more auto parts are made of plastics. Some highway safety experts are concerned that the lighter cars do not provide as much protection for drivers and passengers in case of a collision. The newer cars are also more easily damaged than the heavier old cars with more metal. As a result, parts or even whole cars are more likely to be scrapped. Junked cars are a source of many useful repair parts such as carburetors, mufflers, or tires, and scrap metal is a valuable resource, easily recycled. But plastics are more difficult to recycle effectively, and scrapped cars are now generating millions of pounds of plastic waste each year.

To help solve pollution problems, towns and cities passed regulations against leaf burning. Suddenly, tons and tons of this waste built up all over the country. Valuable landfill space was used up for this "natural" material. Some states, such as New Jersey, reacted to that problem by banning the disposal of leaves in landfills.

Our rising standard of living has resulted in mountains of wastes and pollution of our air, water, and lands. Those problems are likely to increase in the future. Coping with them—no matter what solution we choose—will cost money. Yet few people would suggest that we give up our modern conveniences and return to the horse-and-buggy days. We need careful planning to come up with practical solutions that people will be willing to support.

According to recent polls, the public is willing to pay for measures that will help clean up our environment and deal with our waste glut. They are more willing to put in some personal effort, too. In one survey, 82 percent said they are already voluntarily recycling newspapers, glass, aluminum, and other materials. About 40 million American households are now participating in some kind of curbside recycling program.

Attacking the Trash Crisis

There are four main ways to get rid of garbage: bury it, burn it, recycle it, or make less garbage in the first place. Experts believe that a combination of all four methods is the best way to deal with our trash problems. Such a four-pronged approach is called *integrated waste management.*

We are already using all these techniques to some degree, but our efforts have not usually been coordinated. So far we do not have a national recycling program. The Environmental Protection Agency (EPA) has the power to help states develop solid waste management programs aimed toward conserving resources but cannot force them to adopt specific programs. The federal agency has been attempting to accomplish its aims indirectly, by tightening regulations on landfills and incinerators so that states will have to rely more on recycling programs.

The states do not have coordinated waste management programs,

Curbside recycling pickup in New York City. These newspapers were bound properly, according to the community requirements.

either. Growing numbers of states are adopting recycling laws, but it is up to the individual cities, towns, counties, and other local communities to work out the details of waste management programs. Their approaches vary enormously, even in states where recycling is required by law. Some communities have been dumping all their trash into landfills; some have tried to solve their garbage problems by burning everything in incinerators. Recycling programs vary widely from one town or county to the next, not only in the types of materials but also in procedures. Regulations like not including magazines among bundles of newspapers or sorting glass bottles according to color make sense economically, but the reasons behind them are not always clear to the public. As a result, people who want to do their part to help become angry and resentful when they find the "recyclables" they collected so carefully still piled up by the curb after the scheduled pickup, with a note explaining that they weren't sorted or packaged according to the requirements of the community program.

As for reducing the sources of garbage, we haven't made much progress along those lines. And yet, the Environmental Protection Agency sees that as one of the most important keys to solving the trash problem. The EPA has adopted the philosophy of integrated solid waste management as its blueprint for the future and supports the following plan of action:

1. *Reduce the amount of garbage produced*—by changing our buying habits and redesigning product packaging. Part of the problem is our throwaway attitude. Americans produce twice as much solid waste per person as the people in Europe, where it is a tradition to bring your own string or cloth bag when you go shopping. Europeans often prepare meals with more fresh foods that come without packaging, and paper plates, plastic utensils, individual-sized juice containers with a laminated leakproof lining and attached straw, and "instant" cake mixes complete with a throwaway baking pan have not become popular there. Some American manufacturers, though, have found that reducing the amounts of packaging materials can cut costs. In the past twenty years, soft drink producers have reduced the plastic used in 2-

liter bottles by 21 percent, the aluminum in cans by 35 percent, and the glass in bottles by 43 percent. McDonald's saves 68 million pounds of packaging each year, just by pumping syrup for soft drinks from the delivery trucks into tanks in the restaurants, instead of shipping it in cardboard containers. Fabric softeners are now available in concentrated form that the consumer mixes with water at home, as an alternative to the heavy (and wasteful) throwaway plastic jugs of premixed softener. Many offices are now saving paper waste by using the blank side of old computer printouts and photocopying on both sides of a sheet.

2. *Recycle* as much of as many kinds of waste materials as possible. The EPA has set a national goal of 25 percent recycling by 1992, up from the 10 percent in the 1980s. Experts believe that at least 50 percent of the nation's trash could easily be recycled, and as technology improves we could eventually reach 80 percent.

3. *Use modern, low-pollution-producing incinerators* to burn what is left after recycling. Garbage incinerators of the latest designs are actually complex power plants that take burnable organic wastes (such as food wastes, yard wastes, and paper and wood products) and convert their stored energy to electricity. They are 99.99 percent efficient in eliminating pollutants such as toxic gases and particulates (soot). In fact, they burn much cleaner than coal-burning power plants. The toxic ash they produce does present some disposal problems. But methods are being developed to use incinerator ash as the main ingredient in road pavement, or in inexpensive building materials for the construction of homes and offices.

4. *Bury the rest of the trash in leakproof landfills,* with improved designs that introduce air, moisture, or microorganisms to help speed up decay processes. Many of the landfills in existence today are environmentally unsafe. They leak toxic substances into underground water sources and present a health hazard to the surrounding communities. The EPA has ruled that by 1994, all landfills must be lined with clay or plastic to prevent leakage. As for the present-day landfills that don't meet those specifications—one company, Landfill Mining, Inc., has found a way to "recycle" old garbage dumps. It digs up the old dumps,

removing recyclable materials, then lines the dumps to conform to modern requirements. The "recycled" landfills have become safer, and they add to the scarce supply of space where new unrecyclable garbage can be put.

Why Recycle?

It is clear that the EPA and other waste management experts consider recycling the key to our future efforts to deal with the trash crisis. It has a number of advantages—for individuals, for businesses, and for the environment of Spaceship Earth.

1. *Recycling saves resources.* By reusing iron, lead, and other metals, we conserve our natural resources of these materials. Plastics are made from petroleum, another nonrenewable resource that is getting scarcer; recycling plastics saves on the need for the natural raw materials. Recycling paper, cardboard, and wood products saves on tree-cutting. (Half a million trees—a whole forest—have to be cut down to produce each week's Sunday newspapers!)

2. *Recycling saves money.* As we run out of landfill space, and costs for transporting trash increase, recycling makes more and more sense, economically. It cuts down the amount of trash that must be disposed of and turns part of our wastes into a valuable resource that can be sold to manufacturers. In many cases it costs less to produce new goods from recycled materials than it does to produce them from "virgin" raw materials.

3. *Recycling is kinder to the environment.* Recycling cuts down the amount of trash that would have to be burned, producing forms of pollution that might escape into the environment. Moreover, using recycled sources of glass, metals, paper, and other materials in manufacturing generally requires less energy. Generating extra power not only increases costs but produces more pollution from burning fuels. Recycling paper and wood products has some special benefits. Although trees are renewable resources, which can be replaced by new growth, it takes years for one to grow. But trees help to remove pollutants from the air,

soak up carbon dioxide, and generate oxygen. Thus, they help to keep the air we breathe healthier and counteract the greenhouse effect that may change our planet's climate. The big trees that are cut for lumber and paper products can do a lot more to help the environment than the little seedlings that are planted to replace them. Recycling can make a huge difference. If we could recycle all our newspaper, for example, we could save about 250 *million* trees each year!

Early Recycling Efforts

We tend to think of our current trash crisis as a new problem, and recycling seems like an original and "trendy" approach to solving it. Actually, though, both trash problems and recycling have a long history in the United States.

The first serious attempts to deal with the garbage problem date back to the nineteenth century. (Yes, the trash was piling up even back in the horse-and-buggy days. In fact, the horses produced some waste disposal problems of their own.) In the mid-1890s, New York City street cleaning commissioner George A. Waring introduced an ambitious recycling program to pay for part of the costs of the city's waste disposal by selling recovered materials for a profit. In 1898 he set up the city's first recycling plant, where garbage was sorted and reusable materials were reclaimed. Other cities soon followed New York's example. By 1924, up to 83 percent of American cities were separating some items for reuse.

Industry, too, was discovering that recycling could be profitable. In the 1920s Henry Ford developed charcoal briquettes for barbecues as a way to use the scrap wood left over from building his Model Ts. With the wastes produced in making the briquettes, he produced methanol for antifreeze and ketones for paint for his cars.

Recycling got a real boost during World War II, when many important sources of raw materials were cut off. Americans saved tin cans, glass bottles, cooking fat, and other items of value to the war effort. Valuable supplies of copper, aluminum, and other metals were

The "recycle symbol." The top arrow symbolizes the consumer sending the used packaging or products to the recycling plant. The next arrow symbolizes the recycler giving the raw material to the manufacturer, who will make it into a new product; and the third arrow shows the manufacturer giving the new product to the consumer.

saved, and one third of all paper was recycled. Glass bottles were collected, washed, and refilled dozens of times.

This massive recycling program collapsed when the war ended. There didn't seem to be any more need for it. International trade provided all the supplies of metals and other raw materials we needed, and as for trash problems—landfills were plentiful and cheap. Some cities continued to separate some materials for a while. (One of us recalls growing up in Philadelphia in the 1950s, when there were two collections of "garbage" each week, in addition to the weekly pickup of "trash." The "garbage" was mostly food wastes—in general, anything that would start to rot and smell bad if it stayed around—while "trash" was all the rest of the refuse. The separate collections served two purposes. More frequent garbage pickups reduced health hazards, since rotting food attracted rats, roaches, and other disease-spreading pests; and the food wastes were sold to farmers for composting or as a feed supplement for hogs.) But by the mid-1960s, most communities were collecting mixed garbage; separating materials seemed too expensive to be worth the effort.

In the early 1970s, environmentalists began to worry about the world's dwindling natural resources and about the growing pollution problems. Their concern stirred up interest in recycling programs, and the shift in philosophy was reflected in the titles of some key federal laws. The "Solid Waste Disposal Act" of 1965 was replaced in 1970 by the "Resource Recovery Act," and then, in 1976 by the "Resource Conservation and Recovery Act." More than three thousand recycling centers were set up. But most of these local recycling programs disappeared during the economic recession of 1974–75.

With all the advantages of recycling, why did our efforts in the early seventies fail so badly?

The early efforts were not well organized, and they could not provide a reliable source of material. Moreover, nearly all the programs focused on the same kinds of recyclables: newspapers, aluminum, and glass. At first, many communities were able to pay for their programs by selling recyclables. But as more communities joined in, the amount of these materials exploded. There weren't enough buyers for recyclables, and their value dropped. Soon communities had to *pay* to have their recyclables removed. Some couldn't find takers at any price! A few concerned individuals continued their efforts because they believed they were helping to save the environment. But in most of the country, interest in recycling died out.

In the 1980s, as landfills closed and costs for waste disposal began to skyrocket, communities took a new look at recycling. Now it made good economic sense to try to turn trash into something useful. Growing numbers of recycling programs were set up, not only by large and small communities but also by corporations. Del Monte, for example, not only collects up to four tons of computer waste paper a month at its San Francisco headquarters but has also set up a recycling collection site at one of its research and development centers, for employees whose local communities have no recycling program. Ten years ago, corporations recycled practically no paper at all. Today U.S. companies recycle an estimated 200,000 tons of paper each year. The huge Wal-Mart store chain is helping the recycling effort by issuing a green label for all manufactured products that use recyclable materials or take other measures to help the environment.

People all over are becoming more conscious of the need for recycling and willing to do their part. The "recycle" symbol is becoming familiar, as it appears on products from paper bags and cereal boxes to biodegradable plastic bags. EPA statistics indicate that the amount of recycling, after remaining static at around 11 percent of the municipal solid wastes for a decade, is beginning to rise; in 1988 the recovery rate was more than 13 percent. It seems that this time the recycling effort is here to stay.

5 • Collecting Recyclables

IF YOU JOT down a telephone message on the back of an envelope, paint a bright design on an empty coffee can and turn it into a pencil holder, or cut off the worn-out legs of a pair of jeans to convert them to a usable pair of shorts, you are practicing recycling on a small scale. A large-scale recycling program, for a company or a community, requires a great deal more planning and organization.

Recycling can be divided into several stages:

1. Convincing people to separate recyclable materials from the rest of their garbage.
2. Collecting the recyclable materials.
3. Sorting the recyclable materials.
4. Processing the recyclables into forms that can be reused.
5. Finding markets for recyclable materials so that they can be made into new products.

Keeping each of these stages balanced is the hard part, as we shall see later. Problems arise when one of the stages is not prepared to handle the material generated by the stages before it.

How do you persuade people that it is worth the effort to give up their careless, throwaway habits and separate various kinds of recyclables out of their trash? How do you keep the program going once it has gotten started?

Recycling programs have tried numerous variations of education, incentives, and penalties. The growing awareness of pollution and other threats to the environment has helped make people realize that recycling is a sensible thing to do. Rising bills for trash disposal provide some dollars-and-cents motivation.

But the first really successful recycling efforts were started off as a reaction to a different kind of trash problem—the bottles, cans, papers, and other bits of waste that are carelessly thrown away on roadsides, on beaches, and in parks. This refuse spoils people's enjoyment of nature's beauties, is sometimes a danger to both humans and animals, and is a nuisance and an expense to clean up. Anti-litter laws helped, but didn't solve the problem. So in the 1970s and 80s, a number of states passed "bottle bills" that required people to pay an extra amount each time they bought a bottle of beer, soft drink, or some other beverage. The extra fee was a deposit, which would be returned when the person took the empty bottle back to the store. After collection, the used bottles were recycled.

The bottle laws were a tremendous success. Voters supported them, and they did the job. Of course, not everyone bothered to take back their bottles for "just a few cents," but enterprising children and even adults were glad to go around scouring the streets and beaches for discarded bottles. Because of the deposits, thrown-away bottles were no longer trash; they were worth something. In some places, container litter was reduced by as much as 80 percent.

Soon the aluminum industry joined in and became an even bigger success story. Customers could cash in cans for money at supermarkets and other convenient locations. Many communities and organizations collected cans for fund-raisers. Nearly half of all aluminum cans produced since 1980 were recycled.

Some communities set up drop-off centers for newspapers and other recyclables. Then, as more communities became interested in recycling,

Residents in this community separate their solid waste into three main categories—bottles, aluminum cans, and newspapers—and drop it off at this collection center.

curbside pickups were started, and more of the community residents began to participate. (One study found that 10–30 percent of people participate in community drop-off programs, while 70–90 percent take part in recycling when there are curbside pickups.) While some communities were emphasizing ways to make it easy or even profitable for people to recycle, others began to *require* people to take part. By the end of 1989, thirty states had some recycling laws, recycling was mandatory in ten states, and more than five hundred cities had curbside pickups. In some cases the efforts were a bit too successful: As the

1980s came to a close, so many communities were recycling that there were more of some types of recyclables than the market could handle.

What Can Be Recycled

As a nation we recycled only 10 percent of our garbage during the 1980s. Other countries, such as Japan and Germany, recycle as much as 40 to 50 percent of their garbage. The EPA set a goal for the nation of 25 percent recycling by 1992. As part of the proposed plan, incinerator operators would have to separate recyclable materials like glass, metal, paper, and plastic from the refuse before it is burned. This sorting would be done either at the incinerator plant or beforehand through a community sorting program.

Some experts in the industry believed that the 1992 deadline for 25 percent recycling was too optimistic, and in December 1990 administration pressure forced the EPA to pull back from this goal. David Spencer, an engineer who founded WTE Corp. of Bedford, Massachusetts, thinks the year 2000 would be a more realistic estimate for reaching the 25 percent goal. His company buys up recycling ventures that have failed and makes them profitable. (Many recycling programs fail because the people who started them do not understand how much prices drop when supplies increase. Sometimes they lock themselves into contracts that they later cannot fulfill without losing money.) "We are still learning," says Spencer; but growing experience and new technology should eventually make it possible to recycle 50 or even as much as 80 percent of our garbage.

Many different kinds of materials can be recycled—in one Japanese community residents separate their garbage into thirty-two categories! But many American communities that recycle collect only three main recyclables: newspapers, bottles, and aluminum cans. These three items make up a sizeable fraction of our garbage, so even this limited recycling is a big help in reducing the amount of garbage that remains to be disposed of.

Other communities have much more extensive recycling programs.

Magazines, corrugated cardboard, tin/steel cans (this is the only domestic source of tin), and plastic soda bottles and milk jugs are part of some programs. Others also include scrap paper, junk mail, Styrofoam, large household appliances (which are sold to scrap dealers), oil, tires, and flashlight batteries, as well as leaves, grass, and chipped wood for composting.

The Plastic Debate

The hottest issue in recycling today is plastics recycling. There are many different kinds of plastics, and Americans use 15 to 20 billion pounds a year. Some of them can be recycled, but some kinds can't be recycled with current technology. Plastics make up only about 7 percent of the weight of most garbage collections, but their share of the garbage volume is much greater—up to 30 percent or more. (Most plastics are very light—it takes more than 14,000 plastic bottles to weigh a ton, for example.) Only a little more than 1 percent of all plastics produced is recycled, and 95 percent of our plastic wastes end up in landfills. (The rest is incinerated.)

Plastics have many uses in today's society. Look around your house and see how many things are made with plastics, from the TV to the inside of the refrigerator, to toys, tools, cups, plates . . . Plastics are everywhere. A lot of the packaging for things we buy is plastic, too. We throw away 5 million tons of plastic packaging each year. Plastic takes up a lot of space in landfills, and it doesn't break down. Twenty-five communities, including Minneapolis, Minnesota, and the whole state of Florida, have passed laws restricting the use of plastics. Many of these laws have banned polystyrene foam (Styrofoam), used in packing materials, mattresses, and disposable dishes and trays that have become popular in fast food restaurants, school cafeterias, and even homes. In an effort to avoid these bans, the plastics industry is working hard to make plastics recycling easier and more widespread. They are developing the technology to recycle other types of plastics, increasing facilities for recycling plastics, and finding new uses for recycled plastics. NAPCOR, the National Association for Plastic COntainer Recovery, is a nonprofit association of a dozen major plastics manufacturers that help communities set up plastics recycling programs.

Soft drink bottles and milk jugs are the most widely recycled plastics. Wellman, Inc., of Shrewsbury, New Jersey, for example, buys plastic bottles and turns them into nylon fiber for carpeting, polyester filling for parkas, and heavy felt used to stabilize railroad beds. The recycling of Styrofoam is increasing rapidly. Some schools, for example, have their Styrofoam lunch trays and plastic utensils recycled. In some areas plastic bags, plastic wraps, and other kinds of plastics are also being recycled. National Polystyrene Recycling Co., formed in 1989 by a group of eight chemical companies, is recycling plastics into raw materials for Rubbermaid, Inc. and other plastics manufacturers. It makes good sense, economically, to reuse these materials—recycled products are much cheaper than plastics produced from "virgin" materials.

But there are many pitfalls that make plastics more difficult to recycle than other materials. Their large volume can make curbside collection difficult and expensive; many communities with curbside pickup programs for other recyclables rely on voluntary recycling of plastics at local drop-off centers. Another problem is that there are many different kinds of plastics, each of which must be processed in a different way. Sorting is essential, yet most people can't tell one kind of plastic from another. The plastics industry has proposed a labeling system that many manufacturers voluntarily put on the bottom of plastic containers. At least a dozen states have begun using this coding system, which makes sorting easier both at home and in recycling plants.

If you check some plastic bottles, jugs, or microwave food containers, you will probably find a recycling symbol containing a number, with the abbreviation for the type of plastic under it. Here's how to read the code:

1	PETE	polyethylene terephthalate
2	HDPE	high-density polyethylene
3	V	vinyl
4	LDPE	low-density polyethylene

5	PP	polypropylene
6	PS	polystyrene
7	OTHER	other, including multilayer

Some environmentalists believe recycling is a step in the right direction, but not the right answer. Plastics should not be used so much in the first place, they say. There are more ecologically sound alternatives. Cardboard packaging, for example, could easily replace plastic packaging in many cases. (Recently AT&T changed over from foam to cardboard packaging for shipping its phones.) Plastics do not break down in landfills, toxic pollutants are created when they are manufactured, and petroleum—the raw material for producing plastics—is a limited resource. Cardboard, on the other hand, is made from trees, a renewable resource. Others argue, however, that cardboard actually takes up more space in landfills, and studies have found that even biodegradable products (those that can be broken down by the action of decay bacteria) decompose very slowly in landfills, because they are packed very tightly and there is not enough of the air and moisture that are needed for the bacteria to work. Moreover, plastics can be a renewable resource if they are recycled.

The clamshell-shaped foam packaging used by McDonald's restaurants has become a symbol of the complex plastic problem. In 1990 McDonald's Corporation announced that it would invest $100 million in developing new markets for recycled plastics. The company's polystyrene containers accounted for an estimated 7 to 8 percent of the 1 billion pounds of foam packaging produced in the United States each year. McDonald's set up a test program in its New England restaurants, providing separate bins for customers to deposit the polystyrene containers, which were then collected for recycling. But later that same year, after conducting a joint study with the Environmental Defense Fund on ways of reducing solid waste, McDonald's suddenly announced that it was abandoning the foam hamburger boxes and switching to paper packaging. The new packaging material greatly reduces the volume of waste, but ironically, it cannot be recycled because it contains some plastic. The manufacturer of the new wrap claims that

The clamshell-shaped polystyrene foam packaging formerly used by McDonald's restaurants.

it can be made biodegradable, but some environmentalists, like Jan Beyca of the National Audubon Society, say the switch from plastic to paper will lead to *more* pollution rather than less.

Some communities require that only plastics that are biodegradable (usually those containing cornstarch, which bacteria can feed on, helping to break up the plastic) or photodegradable (broken down by the action of light) may be used. Degradable plastics have become popular in garbage bags and disposable diapers. They are helping a bit with the litter problem, since degradable plastic bags thrown away carelessly break down and disappear after lying out in the open for a while. But the use of degradable plastics can make the overall trash problem worse. Photodegradable plastics in a landfill do not receive the ultraviolet light they need in order to decompose, and, as we mentioned, even naturally biodegradable things do not break down easily in landfills. In addition, degradable plastics get mixed in with nondegradable varieties in recycling collections. They make the recycled plastics less durable and thus reduce the value of the plastic recyclables.

The Diaper Dilemma

Disposable diapers have become one of the most hotly debated items in our trash problem. The arguments for and against their use provide a good illustration of how complicated environmental questions can be. A generation ago, people used cloth diapers on their babies. When each diaper became wet or soiled, it was placed (after dunking in the toilet bowl, if necessary, to get rid of solid wastes) in a diaper pail for temporary storage. When enough dirty diapers had accumulated, the load was washed—either at home or in a laundromat, or by a commercial diaper service that delivered a new load of clean diapers each week. It seemed to be a very ecological system—the diapers were used and reused until they wore out, and then they made good rags for housecleaning. The only problem was the dirty diapers, which were smelly, messy, a lot of work to clean, and particularly inconvenient when you were traveling.

Most babies in the United States today wear disposable diapers, which consist of layers of absorbent paper and a waterproof plastic cover. Americans currently throw away 18 billion disposable diapers each year, and they make up a shocking 2 percent of all our solid waste. Most of them wind up in landfills, where they not only take up valuable space but may also be a health hazard. The instructions on the diaper package say that solid wastes should be washed out before discarding the diapers. But most people don't bother and just throw the whole mess away. Human body wastes can contain more than one hundred different kinds of intestinal viruses, some of which cause diseases such as polio and hepatitis. A 1975 study found that such viruses can survive up to two weeks in a landfill, although Procter & Gamble (the largest manufacturer of disposable diapers) disputes these findings, and a study in Washington State found no evidence that ground waters around landfills are being contaminated by such germs.

Health hazard or not, more than 3 million tons of disposable diapers used each year use up a lot of renewable and nonrenewable resources. Many environmentalists have urged people to switch back to the old-fashioned cloth diapers. Ecologically, it might seem that this would be a far better choice, but several studies suggest that the decision is not so clear-cut. Although reusable cloth diapers consume less raw materials and create less waste, washing them consumes water and energy to heat the water. Detergents used in the washing may pass into streams and lakes and could have harmful effects on wildlife. If a diaper service is used, the delivery trucks add to the environmental costs by burning fuel and emitting exhaust gases.

Regardless of the environmental debate, it seems likely that many people will continue to use disposable diapers, and voters would resist any effort to ban them. Procter & Gamble is leading the effort to develop ways of dealing with disposable diaper waste. In a test program the company has sponsored in Seattle, Washington, one thousand households place their used disposable diapers (whether they are Procter & Gamble brands or not) in recycled plastic bags, which are picked up at the curbside. The diapers are washed, and materials are separated and sanitized. The pulp is used for cardboard boxes, building insulation, and wallboard. The plastic is used for products including garbage bags, flowerpots, and lumber, and the wastes are composted.

Eventually, virtually everything may be recycled. Even cow manure is being recycled by one firm into methane gas, as well as food for pets and farm animals. In California, the world's first power plant that uses cattle manure for fuel burns nine hundred tons of manure a day and provides enough power to run twenty thousand homes. Four hundred thousand cattle graze nearby, producing a never-ending supply of fuel, which had previously gone to waste.

As recycling gains popularity it will become more widespread, easier, and more efficient. Most communities don't recycle yet, however, or offer only limited recycling. Let's look at a few of the nation's role models for recycling.

Model Recycling Programs

In 1989 the Institute of Local Self-Reliance conducted a survey to find the fifteen best recycling programs in America. Three of the top programs were in New Jersey, including two in Camden County. (Oregon also had three communities in the list.) By 1989 New Jersey was leading the rest of the nation in number of communities and businesses that were recycling. Part of this success is due to the state's commitment to helping its communities start up recycling programs. In 1980, in an effort to reach its goal of reducing solid wastes by 25 percent, New Jersey set up a system of grants and loans to help counties launch recycling programs. In 1987, the state made recycling mandatory.

The leader in the 1989 survey was Camden County's Haddonfield, which collected 50 percent of its wastes for recycling. Haddonfield's recycling program began in 1981 with a drop-off center for newspapers. In 1982 curbside collection for glass and newspapers was started. In 1983 the town purchased a four-compartment recycling trailer with a state recycling grant. In 1984 the landfill used by most of Camden County's towns threatened to close, and other landfills were demanding three times the cost. In 1985 Camden County made recycling mandatory to cut down on the amount of garbage. The next year the county opened a processing plant for all of its towns to use, which sorted recyclable materials and processed them for sale to dealers. This meant Haddonfield residents and those of other Camden County towns could put all their recyclables into a single container which was picked up, and the participation rate skyrocketed. Some experts feel this type of recycling program is the most effective because residents don't have to sort out all the different types of recyclables. By 1989, when many communities were just starting to get into recycling, Haddonfield was recycling telephone books, junk mail, envelopes, grocery bags, books, and household appliances, as well as yard waste for composting. As a further incentive to recycle, residents' garbage is left behind if recyclables are found in it, and they can be fined $100.

What makes the Camden County program so successful is cooperation between the processing plant that separates the garbage and the local industries that "recycle" it. In the area are two paper mills, four waste-paper packing plants, three scrap-iron processors, automobile dismantlers, shredders, a company that processes demolition materials, and a processor of tree stumps and brush.

Seattle, Washington, has also made a success of recycling, using a combination of a volunteer approach and cost incentives. Block captains go around the city's neighborhoods, teaching people how to separate their trash for the program, answering their questions, and listening to their gripes. The block captains are prepared to explain, for example, why the recycling program won't pick up plastic yogurt containers and milk jugs (they are made of the wrong kind of plastic for the Seattle program), why magazines can't be combined with newspapers (clay is used in producing the glossy paper of magazines, and it gums up the newsprint recycling machinery), and why rubber-lined tops and lids can't be mixed with metal cans. Residents' trash-collection bills are figured on a pay-by-the-garbage can basis, determined by how much nonrecyclable garbage they leave out for pickups. Any plastic soda bottles, glass, cans, newspapers, and mixed waste paper are picked up free, as long as they are properly separated. As of mid-1990, Seattle was recycling 37 percent of its solid wastes (the highest among the nation's large cities) and was aiming for a goal of 60 percent recycling by 1998.

Berkeley, California, has started an imaginative incentive program to encourage recycling: an Eco-lottery, in which a household chosen at random can win money if no recyclables are found in its garbage.

Experts believe that central processing facilities are an important key to successful recycling programs. Recycling may not be cost-effective in a small-town program because labor is needed to collect the recyclables, sort them, and process them for sale. But when many communities share a central facility, large-scale recycling becomes more realistic.

In 1989 Rhode Island opened a highly automated processing plant that has become a model for other communities. With only six workers, the plant is able to process eighty tons of mixed recyclables a day. Rhode Island's commitment began in 1986, when it passed the nation's first state legislation requiring recycling. It devised a comprehensive plan to reduce wastes with extensive recycling and incineration for energy. Rhode Island residents are given a twelve-gallon blue box into which glass containers, plastic soft-drink bottles, plastic milk jugs, and aluminum and tin cans are placed. Newspapers are placed on top of the blue box. The recyclables are picked up on trash day.

All the recyclables are brought to a highly automated processing center (a materials-recovery facility or MRF). Newspapers are sent off to be sold, and the rest of the recyclables are dumped onto a conveyor belt. An electromagnet pulls off tin/steel cans which are sent to be shredded. The other materials continue down the conveyor belt and meet a rolling chain curtain. Glass containers push through the curtain, but plastic containers and aluminum cans are not heavy enough to pass through, and they go down a different belt. The glass bottles are separated by workers by color: green, clear, and amber. (Recycling mixed-colored glass would produce an unsatisfactory, muddy-colored product.) The glass is then sold to a glass manufacturer. Meanwhile the plastic bottles and aluminum cans continue down the belt and pass through an electrical field. The aluminum cans become positively charged, and when the field is released they pop off the belt. The plastic bottles that remain are sorted manually into harder milk jugs and softer soft-drink bottles, which are then shredded.

A second MRF is planned for the state. It will also be able to handle office paper, cardboard, and other types of plastic, which officials expect will reduce the state's garbage by 25 percent. Nearly three-quarters of the nonrecyclable materials will eventually be incinerated in three "waste-to-energy" plants, which will produce enough energy for 360,000 homes. The ash and the nonburnable materials combined will take up one-tenth of the landfill space that would have been needed.

6 • How Things Are Recycled

SEPARATING our garbage and leaving it on our curb is only the beginning of recycling. Edward Klein, the former head of the EPA task force on solid waste, says, "Most people think they put out the glass, aluminum, and paper, and they've recycled. In fact, all they've done is separate. Until those commodities are taken somewhere else and used again, you haven't recycled."

Communities sell their collected recyclables to dealers, or directly to mills where they are converted into usable raw materials.

Newspaper Recycling

The first paper plant in America, founded in 1690 near Philadelphia, made paper that was 100 percent recycled! Its raw materials were fibers from cotton and linen rags. In the mid-1800s, though, rags could no longer satisfy the growing demand for paper, and new methods were developed to use wood pulp as a raw material. By the 1930s, most paper was made from virgin materials. Recycling paper became a way

of saving scarce raw materials during World War II, but even at the height of the wartime recycling drives, the paper industry used only 35 percent recycled material. After the war, paper recycling dropped sharply, until the new recycling booms of the seventies and eighties. We are making progress—paper recycling is back up to nearly 33 percent—but so far the main effort has been concentrated on only one kind of paper: newspapers.

In 1989, 5.3 million tons of newspapers—one-third of all the newspapers produced that year—were collected for recycling. Two million tons were converted back into newsprint, and the rest was either turned into products like paperboard, insulation, garden mulch, egg cartons, gameboards, book covers, tissue paper, and even cereal boxes, or it was exported. Recycled paper is a very important source for many countries in Asia, Europe, and South America that do not have much timber. In fact, waste paper is becoming one of the largest export items to leave New York harbor.

Only eight newsprint paper mills in the United States and one in Canada use recycled materials. Southeast Paper Manufacturing has set up a state-of-the-art recycling mill in Georgia. Every step of the paper recycling process occurs in one place, and each step uses the latest technology available.

The newspapers are first shoveled onto a conveyor belt, which travels to the pulping machine. There the paper is mixed with water and chemicals and is ground up into pulp. The thickness of the pulpy mixture is measured electronically. If it isn't thick enough, the conveyor belt speeds up the amount of paper flowing into the pulper. If it's too thick, the flow slows down.

The pulpy mixture then goes through several more machines. (One is a cleaner that spins the mixture at high speeds.) Then the pulp passes through a series of screens with narrower and narrower mesh. Contaminants are gradually screened out. Only paper fiber passes through the final screen, whose slots are just 0.008 of an inch wide.

Next the pulp is sent through hollow cylinders with fine-mesh screen walls for de-inking. The pulp clings to the outsides of the cylinders while the ink and water go into the center. The paper is next treated with bleaching chemicals to make it brighter. Then the pulp goes through refiners, which strengthen the cellulose fibers by roughing them with rotating disks. This makes them stick together better (like Velcro fasteners). The pulp is then pressed out in paper machines and through machines that cut and roll the paper.

Paper recycling uses a lot of water—19 million gallons each day at Southeast's mill. (But this is only half the amount of water used to produce paper from virgin timber.) After the paper is produced, sludge is left over. In the past, the sludge was sprayed out on land, but now the moisture is squeezed out, and the dried-out sludge is used as fuel. Special devices remove most of the pollutants when the sludge is burned.

Much of the process is computerized at Southeast's plant. Eventually the company hopes to have every aspect of the process regulated by computers so that someday with just the press of a button the whole process, from start to finish, will go by itself.

Bottle Recycling

The average family uses 207 pounds of glass each year. Unlike paper, which can be recycled only a certain number of times before the fibers break down, glass can be recycled again and again. Unfortunately, though, much of the glass produced never gets recycled. Many bottles, jars, light bulbs, and broken windowpanes are thrown out, and some glass ends up shattered in parking lots and roads when it is carelessly tossed away.

Bottles should be cleaned before putting them out for recycling. Many people are very conscientious about washing out their bottles and removing the labels. But others throw half-full jelly jars into the recycling bin. This makes more work for workers at a processing center. Glass mills need to be sure of the quality of the glass to be encouraged to use recycled glass instead of raw materials.

At a processing center the glass is separated by color—clear, amber,

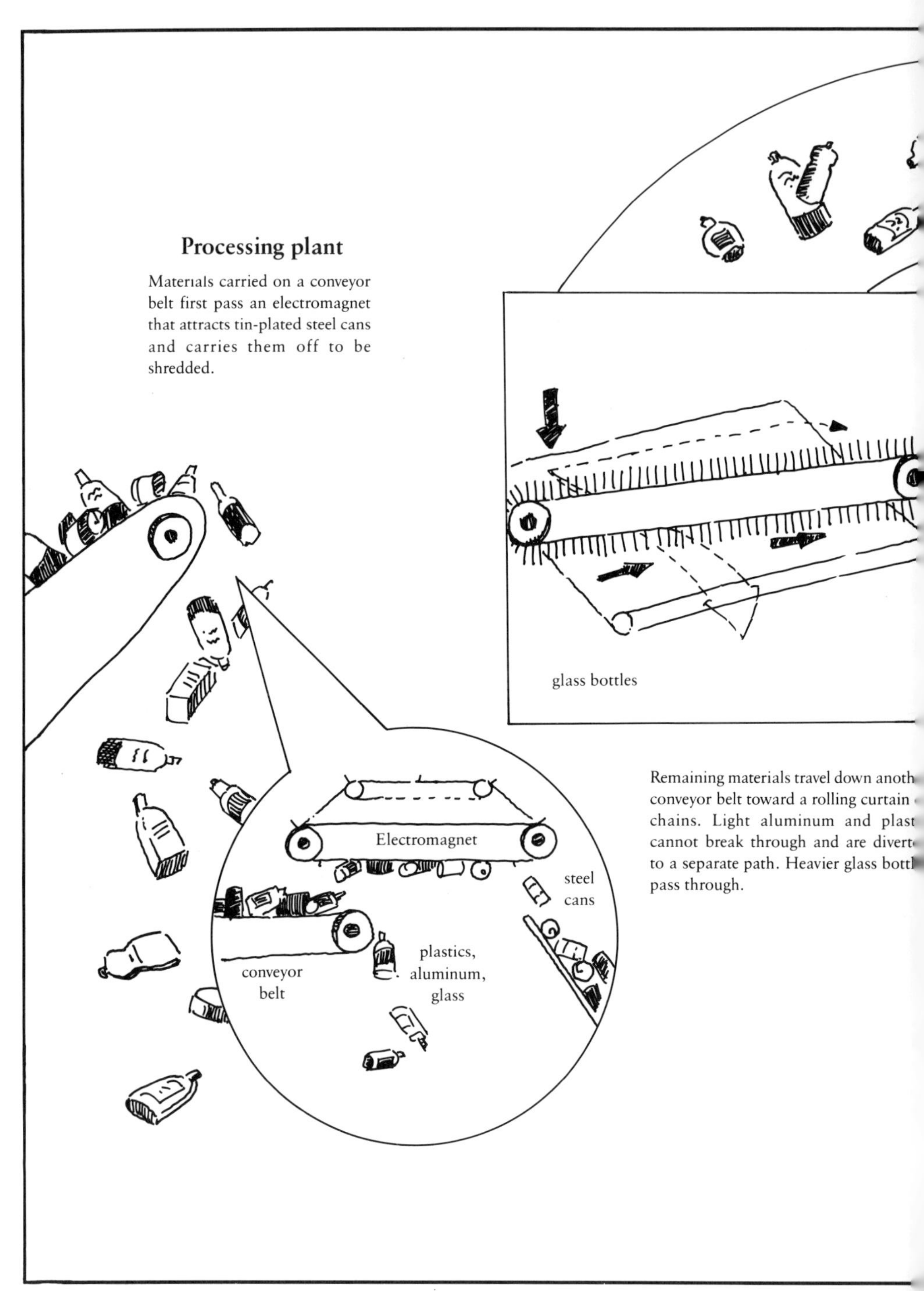

At this materials recovery facility, the aluminum, plastic, and glass are sorted automatically as each item travels down the conveyor belt.

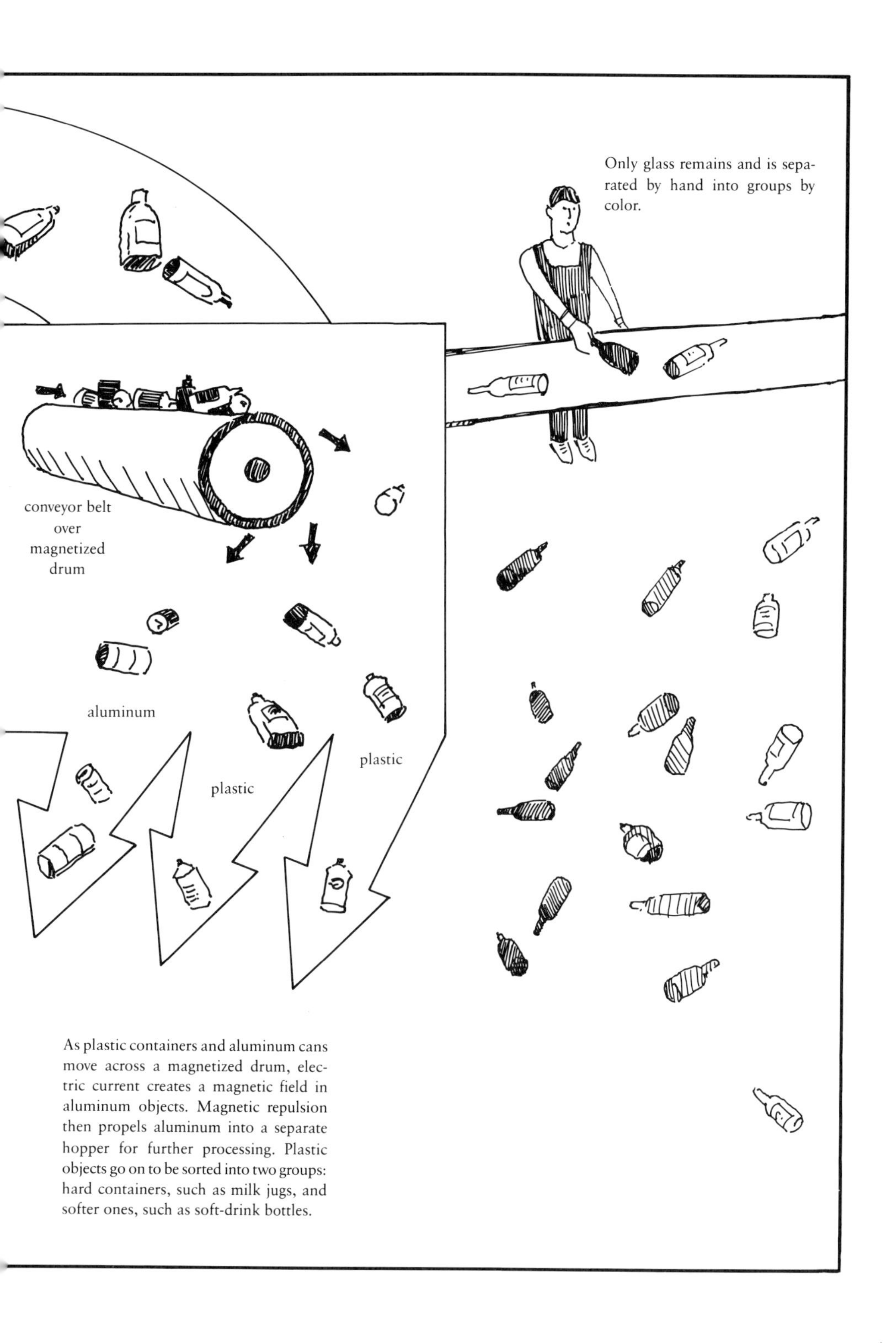

As plastic containers and aluminum cans move across a magnetized drum, electric current creates a magnetic field in aluminum objects. Magnetic repulsion then propels aluminum into a separate hopper for further processing. Plastic objects go on to be sorted into two groups: hard containers, such as milk jugs, and softer ones, such as soft-drink bottles.

green, and mixed—and placed in large bins. These bins are sent to a glass smelter, where the color-separated glass is ground into tiny pieces. These glass particles are heated at high temperatures and are then molded or blown into new glass containers.

In the mid-1970s, less than 20 percent of glass containers were made with the ground glass, or *cullet*. Some glass mills now use 80 to 90 percent cullet when it is clean and free of contaminants. Using cullet actually saves industry money. Sand, limestone, and soda ash—the main components of glass—aren't needed when cullet is used. And cullet melts at lower temperatures than the raw glass materials, so less energy is used to manufacture recycled glass containers. Using less energy saves money, and it also helps the environment.

Aluminum Cans

Aluminum recycling is probably the biggest success story of all, so far. Raw aluminum comes from bauxite ore. When aluminum cans are recycled the process requires 95 percent less energy than producing the cans from raw materials, thus making the finished product cheaper. This is one of the reasons can manufacturers have supported aluminum recycling for many years. More than 300 billion beverage cans were sold between 1981 and 1989. At least half of them have been recycled, including 61 percent of the 80 billion aluminum beverage cans produced in 1989. The recycling of aluminum is so efficient that, according to an estimate by Worldwatch Institute, the average soda can is back on the shelf in only six weeks!

Aluminum dealers were collecting aluminum cans outside supermarkets years before communities began curbside pickups; the cans are now being collected at more than 10,000 recycling centers in the United States. The cans are squashed and sold to an aluminum dealer, who compresses them into huge bales. The bales are then heated at high temperatures. Paint and dirt are separated from the metal, and the aluminum is made into slabs called ingots. These ingots are sold to a can manufacturer, who again melts the aluminum to temperatures of more

Cans begin their recycling journey on a magnetic separating conveyor belt, where they are inspected and where stray steel cans are automatically removed.

than 660° centigrade and pours the molten metal into can molds. The cans are cooled, and the pop-top lids are attached.

Plastics Recycling

Up to now, plastics recycling technology has not come as far as glass and paper recycling. The recycling process for plastics isn't as easy as for other recyclables. One reason is that there are so many different types of plastics, and some are more recyclable than others. Most recycling programs that do include plastics collect only soft-drink containers and milk jugs. Many communities avoid even these plastics, because they are so light and take up a lot of space. (It takes 14,000 plastic bottles to make a ton.) Dealers pay by weight, so for the amount of space plastics take up, recycling doesn't pay very much. It can cost much more to collect the plastics than the community will get for them. Trucks have to unload more often when picking up plastics at the curb, and someone has to sort out the different types of plastics.

When plastic soda bottles were introduced in 1978, nine states quickly introduced bottle bills requiring a deposit on the plastic bottles to keep them out of landfills. Recycling technology for this type of plastic (polyethylene terephthalate, or PET) is much more advanced than for other plastics simply because industry has been recycling it longer. PET plastic is recycled into numerous products, including carpets, fiberfill for coats and sleeping bags, scouring pads and other household products, road barriers, tennis-ball felt, twine, and car parts such as distributor caps.

Milk jugs are made from a harder type of plastic, called HDPE (high-density polyethylene). This is the other main recycled plastic, and it is recycled into plastic pipes, crates, pallets, and plastic lumber, which is used in boat docks, decking, park benches, and fences. (Plastic lumber can also be made from mixed resins.) Polystyrene foam, or Styrofoam, is also being recycled more and more into products like trays, home and office products, insulation board, cassette boxes, Rolodex file holders, and even yo-yos. The FDA restricts recycled plastic products from

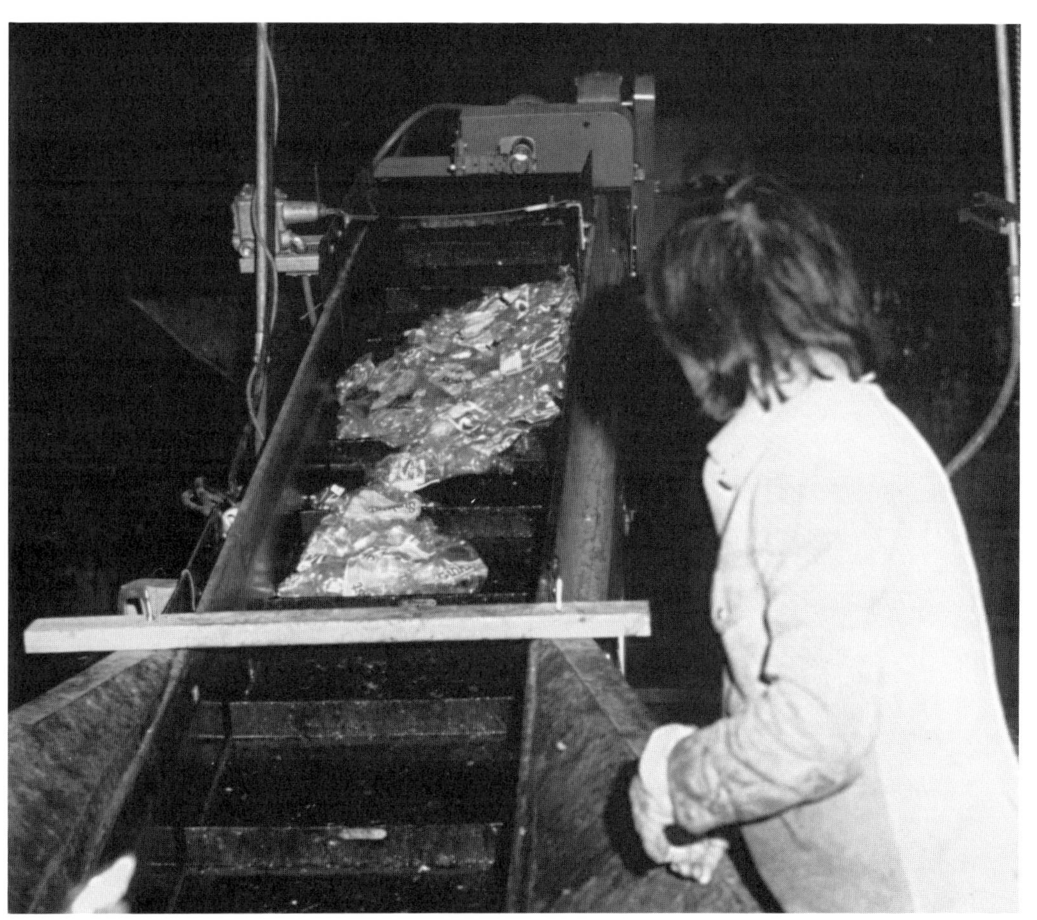

These plastic soda bottles are on their way to be shredded and recycled into other products.

being used as food product packaging, or products used with food. The federal agency claims that the temperature required to melt the plastics during recycling is not high enough to destroy harmful germs and other contaminants. Experiments with layered plastics may make it possible to reuse plastics for food packaging.

The technology for better ways of recycling these and other types of plastics is developing rapidly. The Center for Plastics Research at Rutgers University has developed an automated system that converts plastic soft-drink bottles and milk jugs into raw materials in two ways. One is to recycle the plastics back into their original raw materials, called resins. The other is to convert plastic containers into plastic lumber.

The plastic bottles are sorted by hand into the two main types. Milk, juice, and detergent bottles are mostly HDPE. Plastic soda bottles are made of PET. Both types of bottles are shredded and chopped into fingernail-sized bits.

When the bottles are shredded, there may be contaminants. Many soda bottles, for example, have a hard base made of HDPE. There are also plastic or paper labels and the glue that attached them, as well as aluminum caps.

HDPE milk jug chips (contaminants and all) are poured into a machine that melts and mixes them up. The melted plastic is poured into a mold that rotates and is cooled down in water, producing plastic lumber. Plastic lumber looks like wood and can be sawed, drilled, nailed, and glued like wood.

Soda bottle chips are made into PET resins. They are placed in a giant dishwasher-type machine that washes off glue and labels. (These contaminants go into the lumber-making machine.) The chips then go through a spin cycle where more contaminants are trapped in filters, and then through a rinse cycle. HDPE is lighter than PET and floats to the top, where it is removed. Then the batch is placed in a giant dryer for three hours. Finally it passes into a static electricity generator. The plastic resins cling to the drum, while aluminum contaminants fall into a box. The plastic resins are brushed off. When the resins are converted into pellets, it is almost impossible to tell them apart from virgin resins.

This new system will be used by a number of communities in the United States and abroad. Rutgers researcher Darrell Morrow comments, "There is a tremendous opportunity to absorb just about any plastics we can collect."

Composting Biodegradables

More than half of the solid wastes that currently go into landfills are biodegradable substances—paper, food, cotton and wool clothing, and garden waste. All these could be recycled by natural processes, if the landfills were not packed so tightly. If air and light could get to them, these biodegradables would be converted by decay microbes into compost, a dark, soil-like substance that makes a wonderful fertilizer for farms and gardens.

Even if landfills could be built more efficiently, though, the composting process would still be very slow, and the product would not be very useful. The decomposed organic matter would be mixed with nondegradable materials, some of them toxic. So some presorting is necessary for a successful composting operation. Still, it is worth the trouble considering that it could reduce the trash volume by more than 50 percent.

About four hundred composting operations are currently in operation in the United States. Most of them treat either sewage or garden wastes (but not both—treating sewage is a more complex problem because it can contain disease germs), and the process takes up to six months. But Bedminster Bioconversion Corporation of Cherry Hill, New Jersey, is using new technology to make composting much more efficient. First bulky items like furniture and appliances are removed from the mixed trash. Some of the paper and other burnables are sent to a local utility for fuel, and huge magnets pull out batteries, steel cans, and other ferrous metal objects. Then the trash is mixed with a nitrogen-rich fertilizer and loaded into huge pipes called digestors. The digestors rotate while the action of decay bacteria heats up their contents to 140°F. After a few days the digested mixture is screened to remove nondegradables and taken outside to "cure" in huge piles for several months. During this time, any harmful microorganisms are killed, and the mixture begins to look (and smell) like rich garden soil.

Using the Bedminster method, a facility run by Recomp Inc. in St. Cloud, Minnesota, has demonstrated that the system can also handle disposable diapers. The inner linings of the diapers are broken down in the digestor, and the plastic outer shells are removed when the compost is sifted. They can then be sent to a plastics recycling facility or buried in a landfill.

Depending on how thoroughly composted garbage is digested and screened, it can be used in various ways. Added to the soil of farms or gardens, it provides a source of organic matter, helps the soil retain water, and cuts soil erosion. The nursery and landscaping industries are a huge potential market. Land reclamation, restoring strip mines, highway construction sites, and old landfills, could use enormous amounts of compost. (Its rich supply of nutrients would speed up the growth of grass and other plants on the filled-in areas.) With the new, high-tech methods, says *Fortune* magazine, composting—"nature's way of recycling"—could be the "sleeper" of the recycling industry.

7 • Overcoming Recycling Problems

IN THE PAST chapters we've seen a number of examples of successful recycling projects. Communities have worked out ways to get their residents to cooperate in the effort. New practices like labeling containers with the type of plastic they contain, to make sorting easier, are helping to solve the plastics problem. But our current recycling programs will fail, just like the efforts in the seventies, unless we find ways to solve another key set of problems.

We have to make sure that there is enough *demand* for the materials that can be recycled. If there is no market for recyclables, then their collecting and processing becomes a huge financial drain rather than a money-saving and profit-making enterprise. Even though the nation's recycling programs are still very limited, in some areas recyclables such as newspapers are flooding the market. Some communities that have just started recycling programs have had to leave out newspaper recycling because no one would buy newspapers from them!

Shortly after New York City began recycling in 1989, it ran into problems. New York recycles eight hundred to nine hundred tons of garbage every day, and over one million families were recycling by the end of 1989. But the city's program wasn't able to handle that huge volume. Processing centers where glass and cans were crushed were filled to capacity. Buyers for the materials became harder and harder to find. The sudden surge of newspapers flooded the market, bringing down the prices old newspapers used to command. By 1990 New York City had to pay brokers $10 a ton to take the newspaper away! Small communities in the Metropolitan area, which had been recycling for years, suddenly found they weren't able to sell their newspapers anymore. In Smithtown, Long Island, for example, half the newspapers collected had to be sent to landfills instead. The problem quickly spread throughout the northeast. In mid-1989 Rhode Island was selling newspapers for $25 a ton, but by 1990 the price had dropped to $2 a ton.

Many more processing plants will have to be built to handle the extra load as community recycling efforts expand, but this will take time. And new markets have to be found—or created—to make the effort worthwhile.

Newspaper Recycling

Why are there only eight newsprint plants in the whole United States that operate with recycled newsprint?

Making newsprint from recycled newspapers and making it from raw materials (trees) are very different processes. It is very expensive for a mill to switch over, and because newsprint is so cheap, most mills can't afford to switch. More mills need to be built, but until there is more demand for paper with recycled content, paper manufacturers will be reluctant to do so.

Most recycled paper is more expensive (although with technological advances, companies like Southeast are able to produce high-quality recycled newsprint at a price comparable to virgin newsprint). The price would come down if there were more demand. But newspaper publishers are reluctant to commit themselves to more newsprint unless the quality and supply are assured. (In 1989 the *Los Angeles Times* used 50 percent recycled paper, the *New York Times* used about 8 percent, and the *Wall Street Journal* used only 1 percent.) So there seems to be a "catch 22," a vicious circle that legislators are trying to break with laws.

Laws proposed in at least fifteen states and in Congress would require newspapers to use more recycled print. In New Jersey, for example, legislators have presented a bill that would require all newspapers printed in the state to increase their use of recycled paper to 45 percent, then 65 percent, and finally 90 percent within just a few years. Florida charges a tax on all newsprint not made with recycled paper, and Connecticut set up a timetable requiring its newspapers to use 40 percent and then 90 percent recycled paper. In New York the newspapers worked out a voluntary arrangement stating that if they did not reach a certain level of recycled newsprint use by a certain date (11 percent by 1992, to 40 percent by 2000), then the state would pass legislation requiring it. In other states, similar deals are being proposed to avoid legislation.

Other Paper Recycling

A similar situation has developed with other types of recycled paper. Corporations have found that they can save money on the high costs of disposing their paper trash by selling it to recyclers. Not only has this practice saved many companies money, but it has even been profitable for some. BankAmerica Corp. makes about $175,000 a year by recycling its paper garbage. In 1988 AT&T earned $360,000 by selling used office paper to paper recyclers and, in addition, saved nearly a million dollars that it would have had to pay to get rid of that paper. MCI Communications has earned money by selling forty-two tons of used computer printouts and ledger sheets to companies that turn the material into toilet paper and paper towels. The Coca-Cola Co. donated the money it made recycling paper, cardboard, and newspapers to local charities.

About 200,000 tons of computer paper are recycled each year. All together, 27.6 million tons of paper were recycled by individuals and businesses in 1989. However, as more become involved in recycling, prices are dropping, making it less profitable and therefore providing less incentive for businesses to recycle.

Again, the problem is finding new markets for the increased supply of materials for recycled paper products. Recycled paper is generally more expensive than virgin paper, especially for quality recycled paper, because there is less demand for it—and there is less demand because it is more expensive. Again we have a "catch 22" situation. Paper mills can only produce more paper, which would lower the price, if there is more demand. Also, in order to assure the mills of a steady supply of waste paper, mills will need individuals to supply waste paper, as well. But the network for collecting waste paper from individuals has not really been set up yet—most communities do not collect waste paper. In 1990 a branch of the Sierra Club spotlighted the problem by threatening to sue the White House for failing to recycle the ton of waste paper its staff produces each day, even though such recycling is required by District of Columbia law. When the story hit the news, Fort Howard Corp., a company that turns recycled paper into products such as toilet paper, offered to collect the Executive Branch's scrap paper.

State and local legislators, as well as the United States Congress, have urged that governments be required to purchase a certain percentage of recycled paper products. Some companies are taking the initiative to increase the paper market, too. Wal-Mart stores and McDonald's Corporation, for example, use recycled paper products, including their annual reports, corporate stationery, business cards, computer paper, and copier paper. Using recycled paper has become the "in thing" to do for many businesses, to show their customers and shareholders that they are concerned about the environment.

Making Other Recycling Markets

All around the country, communities and industry are looking for new markets for materials that are currently recycled, as well as for those that aren't.

Scrap dealers don't like to buy glass cullet that is mixed colors. So New York City found a new use for its waste glass in "glasphalt":

roadpaving in which ground-up glass is substituted for gravel. The ground-up glass is 25 percent cheaper than gravel, so although the city is not getting money for the waste glass, it is saving money overall, while using a material that might have had to go to a landfill.

In the past, companies threw out or burned waste materials left over from producing the products we use. With stricter disposal regulations and skyrocketing disposal costs (some types of toxic trash, such as the concentrated ash from incinerators, may cost more than $2,000 a ton to dump, compared to the average of $27 per ton for the average landfill), companies are eager to look into alternative ways to dispose of wastes. They are finding that many kinds of wastes can be recycled and reused.

Dow Chemical used to burn 150 million pounds of some chemical by-products. Now the company sends these chemicals to its plants around the world, where the waste product is turned into dry-cleaning fluid. The company saves $15 million a year in lower disposal costs and in cheaper raw materials for the dry-cleaning fluid.

Exxon Chemical sells the waste product that comes from making polypropylene to a manufacturer that uses the wastes to make caulking material for mobile homes. Exxon doesn't get a lot of money selling the waste product, but it saves over $8 million a year by not having to incinerate the wastes.

Anheuser-Busch built a composting system to transform its brewery sludge into a fertilizer.

In the 1930s, Western Electric bought a copper recycler, Nassau Metals, so that it could recycle copper telephone lines over and over again. Today, Nassau Metals recycles twenty-five truckloads of telephone equipment trash a day and extracts gold, silver, palladium, zinc, lead, and plastic in addition to the copper.

3M Company recycles precious metals and aluminum from its wastes, sells solvent wastes used in production to other companies to be used for fuel, and sells the trimmings from its diaper tape to a company that processes the plastic tape into pellets that are then used to make coat hangers. Some of the company's other plastic waste products are recycled into garden hoses, spatula handles, auto floor pads, and flowerpots. 3M's materials recovery makes nearly $13 million a year for the company, and saves more than $20 million a year in raw materials and lower disposal costs.

New Markets for Plastics

As the plastics industry improves the technology for recycling different kinds of plastics, it is also urging industries to use more recycled resins. Recycled HDPE and PET resins are already cheaper than virgin resins. As technology improves, other plastics resins will also be competitively priced. All around the world new markets are being found that previously would have used virgin plastics. For example, in West Germany 110 tons of plastic trash (fifty truckloads) were used to construct a noise barrier along a freeway. In Jamestown, Rhode Island, volunteers constructed the first playground in the nation that was made mostly with plastic lumber from recycled milk jugs. Procter & Gamble is using recycled plastics for Tide, Cheer, and Downy bottles. In 1990 McDonald's announced that it would begin buying recycled products to build and remodel its restaurants—about 100 million dollars' worth of recycled materials each year! Each year the giant chain builds or remodels 1,400 stores. Insulation, wallboard, roofing, and furniture are some of the items made with recycled materials that McDonald's will be purchasing.

Recycling—the Wave of the Future

New markets for recycled products are opening up all the time, but some people think the government should help things along. In the past the government has often subsidized the use of virgin materials. Companies have received tax credits to exploit many natural resources such as timber for paper, sand for glass, and bauxite for aluminum. They also received other tax breaks on freight costs and on capital investments, which promoted the use of virgin materials over recycled materials.

Some states have tried to turn the tables to encourage recycling by providing tax incentives for those who use recycled materials, or who purchase recycling equipment. Florida and Wisconsin, for example, give tax exemptions to companies that purchase recycled materials. Incentives like these could greatly increase the demand for recycled materials, which in turn would give more incentive for communities and individuals to recycle as many different things as they can and as much as they can.

8 • The Future of Recycling

TODAY ONLY about 13 percent of our wastes are being recycled. And yet the number of landfills available keeps shrinking. Although incinerators, when properly designed and operated, can be virtually free of polluting emissions, their usefulness is limited and they still produce a residue of ash.

If we are to solve our waste crisis, we must cut down the amount of trash we generate. We also need to raise our recycling rate far beyond today's level.

Achieving such a goal will not be easy. Not only must all American communities cooperate and set up and maintain recycling programs, but such programs must be continued far into the future—until new technology provides better alternatives.

Meanwhile, we need to improve our present recycling methods and develop imaginative new uses and markets for recycled products.

Projects for the Near Future

The nation's first plastics recycling plant capable of handling all kinds of plastics was opened by Union Carbide in 1991, at its Piscataway, New Jersey, complex. This plant can handle plastic bags of all kinds, consumer wraps, bottles of all sizes—just about everything American consumers throw away. As more plants like this are built, plastics will no long present a major problem in our recycling efforts.

Meanwhile, projects aimed at developing new markets for recycled products are developing on many fronts.

The Port Authority of New York and New Jersey is sponsoring an experiment in mixing incinerator ash with asphalt, for use in paving roads. Ash has been used in this way for years in Europe and Japan, but there are still some unanswered questions about what long-term effects the heavy metals and other toxic substances in the ash may have on the environment.

Neutralysis Industries in Australia has developed a method of mixing municipal wastes with clay to produce pellets, which are then processed into blocks resembling concrete—but thirty times lighter. Since 1988 a pilot plant has been producing this new building material, Neutralite, and a full-scale plant will soon be in operation.

When New Jersey made recycling mandatory, newspapers quickly flooded the market, and their value in the paper-recycling market dropped rapidly. In Hunterdon County, New Jersey, farmers are trying out a new use for old newspapers. The paper is chopped up and spread in cow stalls. The newsprint soaks up cow droppings better than straw or hay, and it's much cheaper. After three days, the soaked paper is scooped up, loaded into a manure spreader, and transferred to fields, where it is plowed under as a fertilizer. (The mixture is allowed to stand for a season, to decompose, before the fields are planted.) There were fears that the soil might be contaminated with heavy metals, small amounts of which are present in the ink used to print the newspapers; but tests have shown no signs of the metals.

Old auto tires are a big problem today. Nearly 3 billion of them are currently littering our lands and dumps. Yet the rubber contains valuable chemical resources that could be reclaimed. A research group in Canada has produced oils, coal, and gas from old tires by a process called vacuum pyrolysis (heating up to high temperatures under vacuum). So far, though, the method can't compete effectively with other fuel sources, and it is still cheaper just to throw away the tires. But

These used auto tires were just tossed away, but some of the chemicals in the rubber could be used for other purposes.

changing economics and new technology may make tire recycling more profitable in the future.

General Motors is using another pyrolysis method to convert old plastic car-body panels into a compound that can be used to form new body panels. The pyrolyzed material can also be used in concrete and roofing shingles.

Looking farther into the future, new technologies may revolutionize our present views of recycling.

The Plasma Torch

Westinghouse has developed a device called a Plasma Torch. Material is placed between two electrodes, and an arc of electricity is sent across. The temperature rises to several thousand degrees, converting the material into ions (atoms carrying an electric charge) or a superhot state called plasma. This device is still in the research stage, but it would be a neat way to get rid of toxic wastes. Furthermore, since each ion has its own particular combination of electrical charge and mass, theoretically, it could be selectively removed by magnets. Thus if garbage were placed between the arcs and ionized, it could be separated into its original elements—including valuable metals like gold, silver, copper, and platinum. In one operation, we could solve both the pollution problem and the resource problem.

In Australia, CSIRO, a government research organization, recently announced the development of a "plasma arc furnace" that not only destroys toxic chemical wastes, but generates electricity and heat in the process. Temperatures of 30,000° K are reached during the process—four times hotter than the surface of the sun! Although it is not designed to separate the decomposed toxins into their pure elements, the plasma arc furnace is an important step toward developing a Plasma Torch for recycling mixed trash of all kinds.

Engineering "Superbugs"

There have been a number of big oil spills lately. Huge oil tankers have accidentally released crude oil into the ocean. The oil slicks spread out

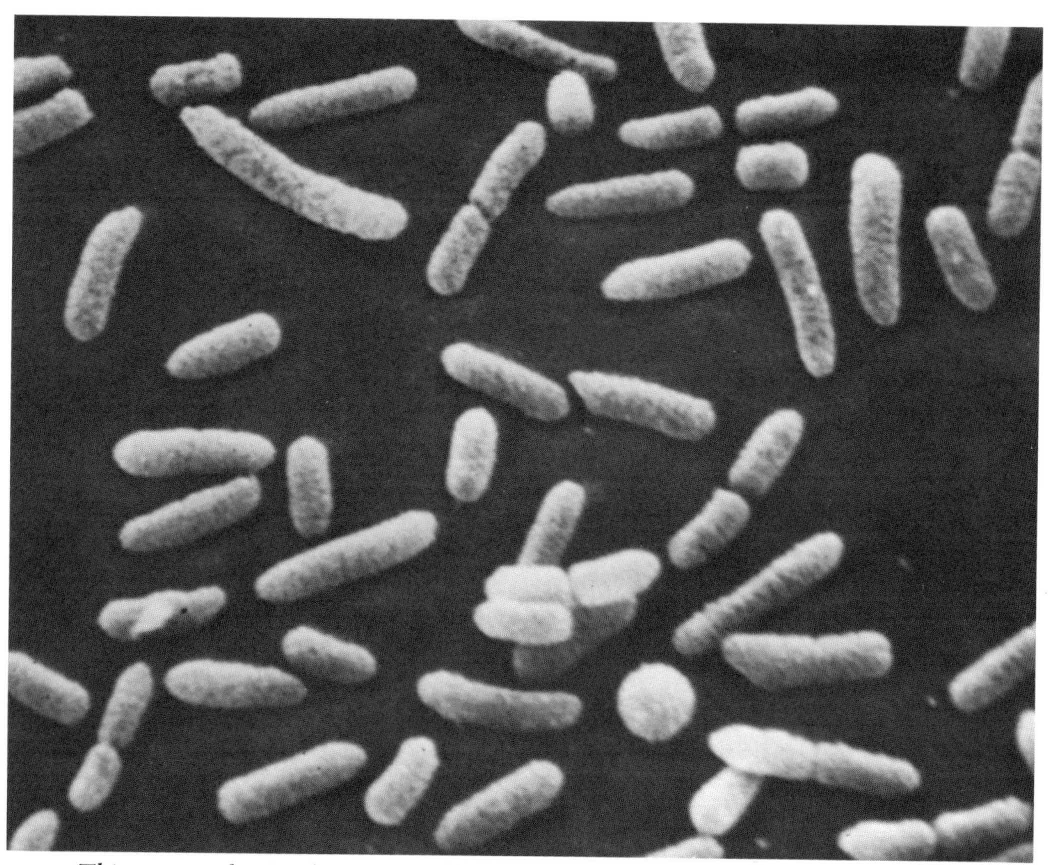

This patented microbe, genetically engineered by General Electric Company, is a "superbug" that actually eats oil in laboratory experiments. It might be used someday to help clean up oil spills in the ocean.

over miles, blackening beaches and killing vulnerable water birds and animals. In some cases it has been possible to clean up the oil and keep it from spreading its damage by adding microorganisms specially bred to have a voracious appetite for oil. The microbes multiply quickly in an oil slick, converting the crude oil to less harmful substances. When the oil is gone, the microbes no longer have a food source and die out. Environmental specialists refer to this approach as "bioremediation."

The first of these "superbugs" was genetically engineered in the laboratory by a research team at General Electric Company, back in 1975. Gene-splicing techniques were used to transfer the hereditary information for an ability to feed on oil from several strains of bacteria to another form. The experiments yielded a hybrid that ate more oil, faster, than any of its "parents." The new microbe was so unusual that its developers were even awarded a patent, just as though they had synthesized a new chemical.

Living creatures have a number of abilities, determined by heredity, that would be very useful in recycling. Some bacteria can digest artificial chemicals—components of plastics, for example. Certain ocean-dwelling animals, plants, and microbes are able to concentrate particular substances out of seawater to an astounding degree. Some animals from the group of bag-shaped bottom dwellers called tunicates, for instance, store the metal vanadium in their body tissues—to such a degree that the vanadium concentration in their bodies is 2 *million* times as great as in the seawater in which they live.

Gene splicers of the future could identify useful genes for abilities like these and transfer them to hardy, fast-growing bacteria. The recycling plants of the future may be huge culture vats in which a sludge of mixed garbage is fermented by genetically engineered decay microbes that break down toxic and presently non-degradable materials and separate out all the useful substances.

9 • What You Can Do

RECYCLING is an important part of the answers to many problems that face us as a society today—pollution, acid rain, the greenhouse effect, the ozone hole, contaminated groundwater, and the lack of places to put our growing amounts of trash.

It is one thing to be concerned about our planet, though, and another to try to do something about it. Many people think that things are so bad, there isn't much any one of us can do to help make the world a better place—so why even try? But people *can* make a difference.

A Changing Attitude

Recycling of glass, aluminum cans, and newspapers is already mandatory in a growing number of states. Eventually it may be required everywhere. Some people have been concerned about recycling for a long time and have done all they could, even when it was quite inconvenient to recycle. Sometimes their neighbors thought they were odd, or fanatics. Today, though, as more people are becoming aware of the problems garbage causes, those who don't recycle are being looked upon as the odd ones, and in some places they are lawbreakers. Recycling is also becoming easier. In many communities, people just have to separate their garbage on trash day.

But there is still a long way to go in the battle for recycling. Many other things besides the three main recyclables can be recycled, but people don't know how to go about it. That situation is changing, though. As concerned people continue to fight for recycling, we are gradually moving toward a completely recyclable society.

The most important step is to change our attitude. Until now we have been a throwaway society, and we just didn't think about what happens to the things we throw away. Now people are discovering that they need to pay more attention. It costs so much to throw things away these days! And more and more, we are beginning to realize that some of the things we throw away so carelessly are endangering our planet—endangering our own lives!

This new attitude doesn't necessarily require us to completely change our lives. It just asks that we be more aware of our daily impact on our world. There are little things each of us can do every day that, together, will make a big difference.

As recycling becomes more popular and spreads through communities across the country, more information is becoming available to let people know what else they can do to help the environment. The 1989 surprise bestseller *50 Simple Things You Can Do to Save the Earth* was followed by a number of other helpful guides for people who are concerned about the future of Spaceship Earth. A list of these resources is found in the "For Further Reading" section at the end of this book.

Precycling

Recycling is only one part of what we need to do to help our planet's environment. It involves finding ways to reuse Earth's resources. Conservation is another important aspect. We need to look at ways to conserve the world's resources so that they will be available when we need them. An important part of conservation is "precycling" or learning to buy and use recycled or recyclable products rather than things that end up in our nation's landfills.

There are many decisions you can make every time you go to the store to buy something. Your family could buy eggs in cardboard containers, for example, rather than Styrofoam, if your community does not have a plastics recycling program. If your supermarket only sells eggs in Styrofoam, write a note telling the manager of the store that you would prefer being able to choose cardboard containers. Buy soft drinks in cans and bottles rather than in plastic containers. Avoid all plastic containers if you can help it. In the course of a year, an average American tears sixty pounds of plastic off products bought in stores. In fact, packaging material accounts for one-third of all the garbage taking up landfill space. Nearly half of cardboard supermarket packages are made from recycled materials. If you're deciding between two brands, choose the one that is wrapped in the smallest amount of packaging material. And send the manufacturer of the other product a note saying that you chose a different brand because of the way it was packaged.

Some markers, paints, and crayons contain toxic chemicals that pollute the environment when you throw them away—or even when you use them. Many crayons, for example, are made from petroleum. Use beeswax crayons instead. Use water-based paints, markers, and glues instead of those containing organic solvents. Use recycled paper to draw or paint on.

One of the biggest precycling issues today is avoiding Styrofoam. When you buy picnic supplies, choose paper plates and cups, not Styrofoam. Lately it seems that everyone is using Styrofoam. Fast-food restaurants and even school cafeterias across the country (except in areas where Styrofoam containers are banned) are using plastic trays, cups, bowls, and utensils. Choose to eat at places that don't use Styrofoam. Write letters to the places that do, and tell them you won't eat there because they offer only Styrofoam containers. The owners want to please their customers. The phaseout of McDonald's Styrofoam clamshell is a good example: If enough people complain, there will be changes.

Some concerned students have won victories against Styrofoam in their schools by convincing School Board members to set up recycling programs for disposable polystyrene trays, cups, and bowls. In New Jersey, students persuaded Board members to switch to paper trays and products, even though they cost slightly more. There is a growing national movement to ban polystyrene, and students at schools around the country are an important part of that movement.

In our throwaway society, "disposable" has almost come to be a synonym for convenience. We need to use fewer disposable products. Instead, find reusable alternatives. Use a rag instead of paper towels, for example, when you need to wipe up a spill; and keep a washable (reusable) cloth towel handy in the kitchen for drying your hands. Reuse things like aluminum foil and plastic bags (rinse them out and let them dry before reusing them). Put away leftovers in reusable containers with lids rather than wrapping them in plastic wrap. Reuse your lunch bag. Ask for paper bags in the store instead of plastic ones, and if you're only getting one or two items, say "No, thanks" when offered a bag. (Although paper bags are recyclable, very few of them contain recycled materials, and most communities don't recycle them. The longer fibers of virgin materials are needed to make grocery bags strong enough to carry heavy loads; but it takes a whole tree, fifteen to twenty years old, to make about seven hundred grocery bags—the number your family might use in a year, if you aren't careful.) Some stores are selling cloth carry-bags that can be used and reused to carry groceries. Often they are printed with sayings like "Save a Tree" or "Save the Planet"—so you can also help to educate people who see you carrying them. Use a mug instead of a Styrofoam cup for hot drinks, and a glass instead of a paper cup for a drink of water.

If there is a baby in your family, rethink the diaper question. Is the convenience of disposable diapers worth adding to the trash load? People who do use them should follow the instructions on the package and wash out messy diapers before discarding them. That will help to stop the spread of disease germs into drinking water. One diaper-changing parent out of six uses cloth diapers, either washing them at home or using a diaper service. Although this is not as convenient, it saves money and landfill space. Lately biodegradable disposables, made with cornstarch, have been attracting some parents. These are an improvement, but not the ideal solution, since the conditions in landfills may not permit even degradable materials to break down completely.

Eliminate Hazardous Household Products

Everyone knows you shouldn't swallow household cleaners, or get them in your eyes, but most people don't think about their effect on the environment. When you throw out containers of oven cleaners, for example, they may not be completely empty. Eventually, the chemicals they contain may leak out of landfills and find their way into our drinking water. Cleansers, insecticides, motor oil, gasoline, paints, paint thinners, turpentine, and stains should all be brought to a household hazardous waste center. But almost everyone just throws them in with the rest of the trash, or pours them down the drain. Only about 10 percent of the hazardous wastes in this country are disposed of properly. According to the EPA, "It has been estimated that in an average city of 100,000 residents, 3.75 tons of toilet bowl cleaner, 13.75 tons of liquid household cleaners, and 3.44 tons of motor oil are discharged into city drains each month." Call the EPA hotline:

(800) 424–9346

to find out where to dispose of hazardous household wastes. A *Household Hazardous Waste Chart* is available from:

Water Pollution Control Federation
601 Wythe Street
Alexandria, VA 22314-1994

For further help in deciding what is the best way to get rid of common household products, look for the 1989 book *Complete Trash: The Best Way to Get Rid of Practically Everything Around the House* by Norm Crampton.

An even better solution to disposing of harmful products is to choose less harmful and more ecological products in the first place. Does your family use detergents with phosphates that can damage lakes and streams? (Liquid detergents don't contain phosphates.) A *Household Hazardous Waste Wheel,* which gives safer alternatives for household products, is available for $3.75 from:

Household Hazardous Waste Wheel
Box 70
Durham, NH 03824-0070

In this handy guide you'll find suggestions such as using homemade cleansers of vinegar, salt, and water, or baking soda and water, or borax and lemon. For furniture polish you can use lemon juice mixed with vegetable oil. For a drain cleaner—baking soda, vinegar, boiling water, and a plunger are suggested. Cedar chips or lavender flowers can be used instead of mothballs; latex or water-based paints are better for the environment than enamel or oil-based paints; soap and water sprayed on leaves, then rinsed off, can substitute for houseplant insecticide; baking soda and powdered sugar can be used as a roach killer. *The Green Consumer,* by John Elkington, Julia Hailes, and Joel Makower, also lists many of these ecologically safer alternatives, as well as others such as mixing two teaspoons of borax and one teaspoon of soap in a quart of water and pouring it into a spray bottle as a general cleaner. The authors suggest sprinkling cornstarch or baking soda on your rug, then vacuuming it up after thirty minutes, as a carpet deodorizer. A half-cup of borax in a gallon of water is an effective disinfectant. Vinegar and water makes a good glass cleaner. In addition to homemade alternatives like these for the home, garden, and much more, *The Green Consumer* goes into great detail about products you can buy that are ecologically safer, with company names and addresses.

Batteries contain chemicals that can cause burns if they leak. They can also contaminate the environment. Many batteries, for example, contain the highly toxic substance mercury. Batteries should never be thrown out with household trash, but that is exactly what happens to most of the 2 billion that are bought every year. Old batteries should be disposed of during "hazardous waste" cleanup days. They can also be recycled, but it's hard to find a place that collects them. Rechargeable batteries cost more than normal batteries, but they can be used over and over again, with a lifetime dozens of times longer than ordinary batteries. Also, when you plug in a tape player instead of running it on batteries, you'll save money and help the environment.

Make Things Last

Don't buy products that you know are poorly made and won't last. Take care of the things you have so that they will last a long time, to cut down on the trash load and save money on replacements. When things are still in good shape but don't fit, or you don't use them anymore, why not donate them to the Salvation Army, Goodwill, or a church bazaar, or even sell them in a garage sale? Someone else will be getting use out of something that would have ended up in the garbage.

Find a Second Use for Things

Look for ways to reuse things you might otherwise throw out. One reader wrote in to *Hints from Heloise* about cutting the sides off a broken laundry basket and converting it into a tray for carrying spillable foods in the car. There are countless "reuses" for almost everything we throw away, from jugs to the cardboard tubes of toilet paper rolls. Cut up the clean parts of junk mail or used looseleaf paper for scrap paper for jotting notes. Offices are finding they can feed the other side of computer paper back into the computer for many jobs that don't require a perfect copy, and photocopy on both sides of a sheet instead of just one.

Don't Throw Out Organic Garbage

Yard waste—leaves and grass—accounts for a large part of our garbage—almost one-fifth. More than half of the trash we throw out each year is organic—food, paper, and other waste products that take up space in landfills. Some people wisely use many of these materials to make rich soil for their gardens or lawns by putting them into a compost pile. Your local agriculture extension service can give you information on setting up a compost pile; or write for a booklet on nontoxic gardens from:

Safer Gardens
P.O. Box 1665
New York, NY 10116

Simply leaving grass clippings on the ground instead of raking them up is an even easier way to recycle garden waste. The cut grass decays and helps to fertilize the lawn.

Save Water

Water is basic to all life on this planet. But 99 percent of the water on the earth is the salt water of oceans and seas, which is not drinkable. Americans use up more than 450 billion gallons of drinkable water every day. Doing your share to conserve water is an important step in the recycling of this vital resource.

Fix leaky faucets and leaky toilet tanks. One out of five American toilet tanks leak. A leaky toilet wastes over 20,000 gallons of water in a year. A slow-drip leak in a faucet can waste more than 3,000 gallons. Don't leave the water running when you brush your teeth. Take short showers instead of baths. Low-flow shower heads will save a lot of water, and you'll never notice the difference. Keep a bottle of water in the refrigerator so you don't have to leave the water running to get a cold drink. Ask your parents to put a clean plastic jug (with some rocks inside to weigh it down) in the toilet tank so that it will use less water when you flush. Five to seven gallons of water are washed down the drain every time you flush the toilet, but with a jug in the tank you can save one to two gallons on each flush. Water lawns in the early morning or evening, when it is less likely to evaporate. Some people collect the rainwater that comes down the gutters of their house in containers that they then use to water plants.

Save Energy

Using less energy will save your family money. It also means the power company has to use less fossil fuels, less nuclear wastes will be produced, and the environment will be better off. Turn lights out when they aren't being used. Dust bulbs, because it will make the room brighter, and they'll last longer and use up less energy. Turn the thermostat down a few degrees in the winter, and set air-conditioner controls up a few degrees in the summer. You'll probably be healthier, and you'll save a lot of energy. If you are going somewhere that isn't too far, ride a bike or walk rather than asking for a car ride. Help your parents make sure your house is properly insulated around the windows and doors. Eliminating drafts saves energy. Keep the refrigerator open as little as possible. The coils in the back of the refrigerator should be dusted to operate efficiently. Keep the temperature inside the refrigerator at the right temperature: not too warm so that things will spoil, but not so cold that ice forms in the milk, for example. About 38–42°F is the recommended temperature for refrigerators. When cooking, cover pots to keep the heat in, keep the oven door closed during baking, and don't use the large-sized burner on the stove to heat up a small pot.

Planting trees in your backyard is another way to save energy. A tree's shade can make your house cooler in the summer, so you won't have to use your air conditioner as much. Planting a tree, growing a garden, or putting plants on your windowsill are also good ways to help the planet by recycling the carbon dioxide in the air. Millions of trees are cut down each Chirstmas, and thrown out when Christmas is over. Buying a live tree and then planting it after Christmas is one way to prolong its life. The International Children's Rainforest Program buys and preserves virgin parkland in South America. Children all around the world have started recycling programs and given the money to the organization.

Find Out About Recycling Other Types of Recyclables

Most communities only accept the main three recyclables at curbside pickups: newspapers, bottles, and aluminum cans. But lots of other kinds of paper products can be recycled, such as cardboard packaging, notepaper, and bags. Most communities accept only aluminum cans, but all kinds of aluminum products can be recycled—aluminum foil, pie pans, and aluminum pots. Tin/steel cans can also be recycled, but usually they aren't allowed in curbside buckets. More and more communities are collecting plastic soda bottles, but most still don't. Find out if there is a place in your area where you can take other recyclable products. Try to start up a program in your neighborhood by letting others know where they can bring recyclables, too. Many people would recycle more if they knew how. To find out about recycling programs in your area, write to:

Environmental Action
1525 New Hampshire Avenue, NW
Washington, DC 20036

Be a Good Example

Don't ever be embarrassed that you care about the planet and recycle. More and more people are realizing it's "cool" to recycle. But some people don't like anything that means extra work, and they might try to make you feel bad, because they don't like feeling guilty about not helping. One of the most important things we can do is to make sure our families and friends recycle where it is mandatory. If your community doesn't have a recycling program, help get one started. Many early recycling programs were set up by individuals who were concerned about their communities. Eventually those programs grew, and recycling became easier for everyone in the community.

Community drop-off centers for recyclables sometimes become places for friendly neighborhood socializing, as well. But the most effective community recycling programs provide curbside pickups. This makes it easiest for people to recycle. Once you set up containers for recyclables—in your garage, for example—recycling becomes a habit and takes less than fifteen minutes of work each week.

Recycling is here to stay. Eventually everyone will recycle, and it won't seem to be extra work, just a normal way of life.

For More Information

The Aluminum Association
900 19th St., NW, Suite 400
Washington, DC 20005

The Council for Solid Waste
Solutions
1275 K Street, NW, Suite 400
Washington, DC 20005

Community Environmental Council
930 Miramonte Drive
Santa Barbara, CA 93109

Earthworm
186 South Street
Boston, MA 02111

Environmental Action
1525 New Hampshire Ave., NW
Washington, DC 20036
(202) 745–4871

Environmental Defense Fund
257 Park Avenue South
New York, NY 10010

Environmental Protection Agency
Office of Solid Waste
401 M Street, SW
Washington, DC 20460
(800) 424–9346

Environmental Task Force
1012 14th St., NW, 15th floor
Washington, DC 20005
(202) 745–4870

INFORM
381 Park Avenue South
New York, NY 10016

Institute for Local Self-Reliance
2425 18th St., NW
Washington, DC 20009
(202) 232–4108

Institute of Scrap Recycling
Industries
1627 K Street, NW—Suite 700
Washington, DC 20006
(202) 466–4050

Keep America Beautiful
Mill River Plaza
9 West Broad Street
Stamford, CT 06902

Local Solutions to Global
Pollution
2121 Bonar Street, Studio A
Berkeley, CA 94702

National Association for Plastic
Container Recovery (NAPCOR)
5024 Parkway Plaza Blvd. #200
Charlotte, NC 28217
(704) 357–3250

National Association of
Recycling Industries
330 Madison Avenue
New York, NY 10017

National Recycling Coalition
(NRC)
1101 30th St., NW, Suite 305
Washington, DC 20007

National Solid Waste Management
Association
1730 Rhode Island Ave., NW
Washington, DC 20036
(202) 659–4613

Paper Recycling Committee
American Paper Institute
260 Madison Ave.
New York, NY 10016

Renew America
1400 16th St., NW, Suite 700
Washington, DC 20036

Steel Can Recycling Institute
680 Andersen Dr.
Foster Plaza 10
Pittsburgh, PA 15220
(800) 876-SCRI

SWICH
(The Solid Waste Information
Clearinghouse)
P.O. Box 7219
Silver Spring, MD 20910
(800) 67-SWICH

A free booklet *Recycling Works!* is available from the EPA Office of Solid Waste at (800) 424–9346.

Glossary

BIODEGRADABLE: capable of being broken down by the action of decay bacteria.

COMPOST: decayed organic matter (food and garden wastes) used as a fertilizer and conditioner for soil.

CORRUGATED CARDBOARD: heavy paper material shaped into parallel ridges and glued together in layers to provide strength and rigidity.

CULLET: scrap glass, usually broken up into small, uniform pieces.

FOOD WASTE: unwanted materials produced in the preparation and production of fruits, vegetables, poultry, meat, and other foods.

GARBAGE: food waste; a synonym for refuse or trash.

LANDFILL: a dumping place for solid wastes. It is lined with clay and impermeable synthetics to prevent leakage into groundwater, and each layer of trash is covered with a layer of soil, then compacted with bulldozers. (If the landfill construction provides for collecting and treating leachate, the layers are not compacted.)

LEACHATE: liquid material that leaks out of a landfill.

MATERIALS RECOVERY: the extraction of useful materials (such as paper, glass, or metals) from solid waste, to be followed by reprocessing for reuse.

MIXED PAPER: waste paper of various kinds of qualities.

PHOTODEGRADABLE: capable of being broken down by the action of light.

POST-CONSUMER WASTE: a material or product that has been discarded after being used for its original purpose. (Newspapers and used paper bags collected for recycling are post-consumer waste but the paper scraps trimmed off during the manufacture of newspapers or bags are not, although they can also be recycled.)

RECYCLABLE: materials and resources that have useful chemical or physical properties after serving a specific purpose and can therefore be reused for the same or other purposes.

RECYCLED: materials which have been extracted from the solid waste stream (or removed before they enter it) and are reused in the original or a changed form.

REFUSE: worthless or useless material; trash.

SLUDGE: mud; sediment; the solid matter produced by sewage treatment.

SOLID WASTE: the unwanted solid material produced by homes, commercial enterprises, government, industry, and agriculture.

SOURCE REDUCTION: decreasing the amount of waste, both by eliminating unnecessary components (including packaging) and by producing more durable, long-lasting, easily repaired goods.

SOURCE SEPARATION: sorting and collection of recyclable materials from trash before it is mixed with the solid waste stream.

TRASH: broken, crumbled, or worthless material; garbage.

WASTE STREAM: all the trash produced by a community. Collecting and disposing of these wastes are usually the responsibility of the local government.

For Further Reading

Books

Blumberg, Louis and Robert Gottlieb. *War on Waste: Can America Win Its Battle with Garbage?* Island, 1989.

Crampton, Norm. *Complete Trash: The Best Way to Get Rid of Practically Everything Around the House.* Evans, 1989.

Earth Works. *Fifty Simple Things You Can Do to Save the Earth.* Earthworks, 1989.

Earth Works. *Fifty Simple Things Kids Can Do to Save the Earth.* Andrews, McMeel, 1990.

Earth Works. *The Recycler's Handbook.* Earthworks, 1990.

Elkington, John and others. *The Green Consumer.* Tilden Press, 1990.

Global Tomorrow Coalition. *The Global Ecology Handbook: What You Can Do About the Environmental Crisis.* Beacon Press, 1990.

MacEachern, Diane. *Save Our Planet: 750 Everyday Ways You Can Help Clean Up the Earth.* Dell, 1990.

Miller, Christina G. and Louise A. Berry. *Wastes.* Watts, 1986.

Naar, Jon. *Design for a Livable Planet: How You Can Help Clean Up the Environment.* Harper & Row, 1990.

Newsday. *Rush to Burn: Solving America's Garbage Crisis.* Island, 1989.

O'Connor, Karen. *Garbage.* Lucent, 1989.
Pringle, Laurence. *Throwing Things Away.* Crowell, 1986.

Articles

Beck, Melinda et al., "Buried Alive," *Newsweek,* November 27, 1989, pp. 66–76.
Brown, Patricia L., "Recycling Group's Winners and Sinners of Product Packaging," *New York Times,* November 9, 1989, p. C12.
Cook, William J., "A Lot of Rubbish," *U.S. News & World Report,* December 25, 1989/January 1, 1990, pp. 60–62.
Goldoftas, Barbara, "Recycling Programs Are Taking Off Around the Country," *Technology Review,* November-December 1987, pp. 29–35, 71.
Lipkin, Richard, "Recycling, King of the Trash Heap," *Insight,* February 26, 1990, pp. 48–49.
Marinelli, Janet, "Composting: From Backyards to Big-Time," *Garbage,* July/August, 1990, pp. 44–51.
Rice, Faye, "Where Will We Put the Trash?" *Fortune,* April 11, 1988, pp. 96–100.
Stevens, William, "When the Trash Leaves the Curb: New Methods Improve Recycling," *New York Times,* May 2, 1989, pp. C1, C6.

Additional articles of interest can be found in the magazines:

Environment. Heldref Publications. 4000 Albermarle Street, NW, Washington, DC 20016 (800) 365–9753.

Garbage: The Practical Journal for the Environment. Old House Journal Corporation. 435 Ninth Street, Brooklyn, NY 11215-9937 (800) 274–9909.

P3: The Earth-based Magazine for Kids. PO Box 32, Montgomery, VT 05470 (802) 326–4669.

INDEX